This is an undergraduate textbook on the physics of electricity, magnetism, and electromagnetic fields and waves. It is written mainly with the physics student in mind, although it will also be of use to students of electrical and electronic engineering. The approach is concise but clear. The authors have assumed that the reader will be familiar with the basic phenomena; however, they have set out the theory in a completely self-contained and coherent way.

After a short mathematical prologue, the theory of electricity and magnetism, and the relationship between them, is developed. The relationship between the microscopic structure of matter and the macroscopic electric and magnetic fields is stressed throughout, and the theory is developed to the point where the reader can appreciate the beauty and coherence of the Maxwell equations which describe these fields. The theory is then applied to a wide range of topics from the properties of materials, including semiconductors and superconductors, to the generation of radiation by electric currents. A final chapter makes the connection between Maxwell's equations and the special theory of relativity. Each chapter ends with a set of problems, answers to which are also provided.

The authors have extensive experience of teaching physics to undergraduate students at the University of Bristol. The clarity of the mathematical treatment they provide will facilitate a thorough grasp of the subject, and makes this a highly attractive text.

Electricity and Magnetism

Electricity and Magnetism

W. N. Cottingham
University of Bristol

D. A. Greenwood
University of Bristol

CAMBRIDGE
UNIVERSITY PRESS

Published by the Press Syndicate of the University of Cambridge
The Pitt Building, Trumpington Street, Cambridge CB2 1RP
40 West 20th Street, New York, NY 10011-4211 USA
10 Stamford Road, Oakleigh, Melbourne 3166, Australia

First published 1991
Reprinted 1995

Printed in Great Britain by Athenæum Press Ltd, Gateshead, Tyne & Wear

British Library cataloguing in publication data
Cottingham, W. N.
Electricity and magnetism.
1. Electromagnetism
I. Title II. Greenwood, D. A.
537

Library of Congress cataloguing in publication data
Cottingham, W. N.
Electricity and magnetism/W. N. Cottingham, D. A. Greenwood.
p. cm.
ISBN 0 521 36229 6. – ISBN 0 521 36803 0 (pbk.)
1. Electromagnetism. I. Greenwood, D. A. II. Title.
QC760.C674 1991
537–dc20 90-26471 CIP

ISBN 0 521 36229 6 hardback
ISBN 0 521 36803 0 paperback

UP

Contents

Contents

Preface

This is an undergraduate textbook on the physics of electricity, magnetism, and electromagnetic fields and waves. It is written mainly with the physics student in mind, although it will also be of use to students of electrical and electronic engineering. We have aimed to produce a concise text which emphasises the meaning and significance of the concepts that appear in the theory, and the overall coherence and beauty of the Maxwell equations.

The theory is set out in a self-contained way, but we assume that the reader will already have some knowledge of the basic phenomena of electricity and magnetism. (At the University of Bristol there is an established tradition of demonstration experiments in the introductory first year physics lectures.) We also assume some familiarity with the mathematics of scalar and vector fields, and the properties of the ∇ operator. The basic theorems are set out for reference in the Mathematical Prologue. The Dirac δ-function is introduced in a non-rigorous way on the first page of Chapter 1, and used freely: in our experience, physics students readily accept this as an obviously useful mathematical device. A few other mathematical tools are developed in the text, as and when they are needed. To avoid impeding the flow of the main argument, the technical details of mathematical manipulations are sometimes relegated to the problems.

The relationship between the microscopic structure of matter and the macroscopic fields which are the main concern of the text is stressed from the start, albeit from a classical standpoint. Not only is this basic for an understanding of the theory, but it is important to appreciate the limitations on the theory's domain of applicability. For example, the important technology of electronic devices lies at the boundary between what is clearly microscopic and what is clearly macroscopic; an appreciation of the nature of the macroscopic fields in materials is essential to an understanding of the underlying physics.

We arrive at the Maxwell equations in Chapter 5, as abstractions from laboratory experiments. Subsequent chapters cover a wide range of applications, from the macroscopic description of the electric and magnetic properties of material media, including superconductors, to the generation of radiation by electric currents. The final chapter makes the connection between the Maxwell equations and the special theory of relativity.

We envisage our book as one to work from. The problems following each chapter illustrate and extend the text, and form an essential part of that work.

We are grateful to Bob Chambers, David Gibbs, Brian Pollard, and many other colleagues, who have helped clarify our presentation of the subject at many points. We thank Margaret James, who worked valiantly on the typescript.

W. N. Cottingham
D. A. Greenwood

Units, constants, and formulae

SI units

	Name	Symbol	Basic units
electric current	ampere	A	A
electric charge	coulomb	C	s A
potential difference	volt	V	$kg\ m^2\ s^{-3}\ A^{-1}$
capacitance	farad	F	$kg^{-1}\ m^{-2}\ s^4\ A^2$
resistance	ohm	Ω	$kg\ m^2\ s^{-3}\ A^{-2}$
magnetic field	tesla	T	$kg\ s^{-2}\ A^{-1}$
magnetic flux	weber	Wb	$kg\ m^2\ s^{-2}\ A^{-1}$
inductance	henry	H	$kg\ m^2\ s^{-2}\ A^{-2}$

$$\varepsilon_0 \approx 8.85 \times 10^{-12}\ kg^{-1}\ m^{-3}\ s^4\ A^2$$

$$\mu_0 = 4\pi \times 10^{-7}\ kg\ m\ s^{-2}\ A^{-2}$$

$$(\mu_0 \varepsilon_0)^{-1} = c^2, \quad c = 299\,792\,458\ m\ s^{-1}$$

Physical constants

Proton charge	e	1.60218×10^{-19} C
Electron mass	m_e	9.1094×10^{-31} kg
Proton mass	m_p	1.67262×10^{-27} kg
Boltzmann's constant	k_B	1.3807×10^{-23} J K^{-1}
Avogadro's number	N_A	6.0221×10^{23} mol^{-1}
Planck's constant	h	6.6261×10^{-34} J s

$$1\ eV \approx 1.60218 \times 10^{-19}\ J$$

Notation

$\mathbf{r}$, $\mathbf{k}$, etc., denote vectors (x, y, z), (k_x, k_y, k_z), and $r = |\mathbf{r}| = (x^2 + y^2 + z^2)^{\frac{1}{2}}$, $\hat{\mathbf{r}} = \mathbf{r}/r$, etc.

Spherical polar coordinates are denoted by (r, θ, ϕ).

Cylindrical polar coordinates are denoted by (ρ, ϕ, z), where $\rho = (x^2 + y^2)^{\frac{1}{2}}$.

$$\nabla^2 = \frac{\partial^2}{\partial x^2} + \frac{\partial^2}{\partial y^2} + \frac{\partial^2}{\partial z^2} = \frac{1}{r}\frac{\partial^2}{\partial r^2}r + \frac{1}{r^2 \sin\theta}\frac{\partial}{\partial\theta}\sin\theta\frac{\partial}{\partial\theta} + \frac{1}{r^2 \sin^2\theta}\frac{\partial^2}{\partial\phi^2}.$$

Vector identities

u, v denote scalar functions; $\mathbf{F}$, $\mathbf{G}$ denote vector functions.

$$\nabla = \mathbf{i}\frac{\partial}{\partial x} + \mathbf{j}\frac{\partial}{\partial y} + \mathbf{k}\frac{\partial}{\partial z}$$

$$\nabla u = \left(\frac{\partial u}{\partial x}, \frac{\partial u}{\partial y}, \frac{\partial u}{\partial z}\right) = \frac{\partial u}{\partial x}\mathbf{i} + \frac{\partial u}{\partial y}\mathbf{j} + \frac{\partial u}{\partial z}\mathbf{k}$$

$$\nabla \cdot \mathbf{F} = \frac{\partial F_x}{\partial x} + \frac{\partial F_y}{\partial y} + \frac{\partial F_z}{\partial z}, \quad \nabla \times \mathbf{F} = \left(\frac{\partial F_z}{\partial y} - \frac{\partial F_y}{\partial z}, \frac{\partial F_x}{\partial z} - \frac{\partial F_z}{\partial x}, \frac{\partial F_y}{\partial x} - \frac{\partial F_x}{\partial y}\right)$$

$$\nabla \cdot (\nabla \times \mathbf{F}) = 0, \quad \nabla \times \nabla u = 0$$

$$\nabla \times (\nabla \times \mathbf{F}) = \nabla(\nabla \cdot \mathbf{F}) - \nabla^2\mathbf{F}, \text{ where } \nabla^2\mathbf{F} = (\nabla^2 F_x, \nabla^2 F_y, \nabla^2 F_z)$$

$$\nabla(uv) = u\nabla v + v\nabla u$$

$$\nabla \cdot (u\mathbf{F}) = \mathbf{F} \cdot \nabla u + u\nabla \cdot \mathbf{F}$$

$$\nabla \cdot (\mathbf{F} \times \mathbf{G}) = \mathbf{G} \cdot (\nabla \times \mathbf{F}) - \mathbf{F} \cdot (\nabla \times \mathbf{G})$$

$$\nabla \times (u\mathbf{F}) = \nabla u \times \mathbf{F} + u\nabla \times \mathbf{F}$$

$$\nabla \times (\mathbf{F} \times \mathbf{G}) = \mathbf{F}(\nabla \cdot \mathbf{G}) - \mathbf{G}(\nabla \cdot \mathbf{F}) + (\mathbf{G} \cdot \nabla)\mathbf{F} - (\mathbf{F} \cdot \nabla)\mathbf{G}$$

Useful results: $\nabla \cdot \mathbf{r} = 3$, $\quad \nabla(r^n) = nr^{n-1}\hat{\mathbf{r}}$

If $\mathbf{a}$ is a constant vector, $\nabla(\mathbf{a} \cdot \mathbf{r}) = \mathbf{a}$, $\quad (\mathbf{a} \cdot \nabla)\mathbf{r} = \mathbf{a}$.

Maxwell's equations

$$\mathbf{\nabla} \cdot \mathbf{E} = \rho / \varepsilon_0$$

$$\mathbf{\nabla} \times \mathbf{B} - \mu_0 \varepsilon_0 \frac{\partial \mathbf{E}}{\partial t} = \mu_0 \mathbf{J}$$

$$\mathbf{\nabla} \cdot \mathbf{B} = 0$$

$$\mathbf{\nabla} \times \mathbf{E} + \frac{\partial \mathbf{B}}{\partial t} = 0$$

In material media, these become

$$\mathbf{\nabla} \cdot \mathbf{D} = \rho_{\text{free}}$$

$$\mathbf{\nabla} \times \mathbf{H} - \frac{\partial \mathbf{D}}{\partial t} = \mathbf{J}_{\text{free}}$$

$$\mathbf{\nabla} \cdot \mathbf{B} = 0$$

$$\mathbf{\nabla} \times \mathbf{E} + \frac{\partial \mathbf{B}}{\partial t} = 0$$

where

$$\mathbf{D} = \varepsilon_0 \mathbf{E} + \mathbf{P},$$

$$\mathbf{H} = (\mathbf{B}/\mu_0) - \mathbf{M}$$

Glossary of symbols

$\mathbf{A}(\mathbf{r})$, $\mathbf{A}(\mathbf{r}, t)$ vector potential
a   acceleration
a_0 Bohr radius
B   magnetic field
$\mathbf{B}_c$, B_{c1}, B_{c2} critical magnetic fields
C capacitance
$\mathbf{D} = \varepsilon_0 \mathbf{E} + \mathbf{P}$ electric displacement
$\mathbf{E}$, $\mathbf{E}_{at}$ macroscopic, atomic electric field
$\mathscr{E}$ electromotive force (e.m.f.), energy
F   force
$F(r)$ §1.2 averaging function
$\mathscr{F}$ magnetic flux
$\mathbf{H} = (\mathbf{B}/\mu_0) - \mathbf{M}$
I electric current
$\mathbf{J}$, $\mathbf{J}_{at}$ macroscopic, atomic current density
$K = 2\omega\kappa/c$ absorption coefficient
k   wavevector, $|\mathbf{k}| = 2\pi/\lambda$
L   angular momentum vector
L   length, §1.2 averaging length
L, L_{ij} self, mutual inductance
dl   line element
M   magnetisation
m   magnetic dipole moment
N   Poynting vector
N number per unit volume
n̂   unit normal to surface
$n = \mathrm{Re}(\sqrt{\varepsilon_r})$ refractive index
P   polarisation
P power
p   electric dipole moment

Q electric charge
Q_{ij} electric quadrupole tensor
R resistance, §12.7 reflection coefficient
$d\mathbf{S}$ element of surface
S surface area
T temperature, §12.7 transmission coefficient
T_c critical temperature
U field energy
V volume, potential difference
W work function
Z_0 characteristic impedance

α atomic polarisability
γ damping constant
δ skin depth
$\varepsilon_r = 1 + \chi_e$ relative permittivity (dielectric constant)
$\kappa = \mathrm{Im}(\sqrt{\varepsilon_r})$
Λ London penetration depth
λ wavelength
μ magnetic moment of molecule
$\mu_r = 1 + \chi_r$ relative permeability
ν frequency
$\rho, \rho_{el}, \rho_{at}$ macroscopic, electronic, atomic charge density
σ surface charge density, electrical conductivity, cross-section
τ collision time, damping time
$\Phi(\mathbf{r}), \Phi_{at}(\mathbf{r})$ macroscopic, atomic electrostatic potential
$\Phi(\mathbf{r}, t)$ scalar potential
Φ_0 §14.5 flux quantum
ϕ angular coordinate, phase angle
$\chi(\mathbf{r}), \chi(\mathbf{r}, t)$ §9.1, §16.1 gauge function
χ_e electric susceptibility
χ_m magnetic susceptibility
$d\Omega$ element of solid angle
ω angular frequency
ω_p plasma frequency

Mathematical Prologue

In the chapters which follow, we assume that you are already familiar with the basic mathematics of scalar and vector fields in three dimensions, the properties of the ∇ operator, the integral theorems which hold for these fields, and so forth. In this prologue, we remind you of some basic definitions, and outline (without proof) those mathematical theorems of which we shall make extensive use. We also establish our notation and sign conventions.

We envisage space filled with electromagnetic fields, and at any instant we describe these fields mathematically using functions which may be scalar functions of position (like the potential $\Phi(\mathbf{r})$) or vector functions of position (like the electric field $\mathbf{E}(\mathbf{r})$). We shall assume that the functions which appear in the theory are continuous, and have derivatives existing as required, except perhaps at special points or on special surfaces. Singularities in the mathematics will usually correspond to singularities in the physics. For example, the electrostatic potential of a point charge Q at the origin is $Q/4\pi\varepsilon_0 r$, and this function satisfies our conditions except at $\mathbf{r} = 0$, which is the position of the point charge.

We sometimes focus on these fields in limited regions of space, say inside a volume V enclosed by a surface S, or over a surface $S(\Gamma)$ bounded by a curve Γ.

P.1 Volume integrals

Volume integrals will often arise naturally in the theory, for example when we calculate the total charge or total energy in some volume V of space. The volume integral of a function $f(\mathbf{r})$ is defined by

$$\int_V f(\mathbf{r})\,\mathrm{d}V = \mathrm{limit}\,(\sum_i f(\mathbf{r}_i)\,\delta V_i), \qquad (\text{P.1})$$

where the volume V is dissected into elements δV_i, and $\mathbf{r}_i$ lies in δV_i. The linear dimensions of the δV_i go to zero in the limit.

1

We might for example take $dV = dx\,dy\,dz$ so that

$$\int_V f(\mathbf{r})\,dV = \int\int\int f(x, y, z)\,dx\,dy\,dz,$$

or work in spherical polar coordinates, and take $dV = r^2 \sin\theta\,dr\,d\theta\,d\phi$, or use any other coordinate system which is convenient for the problem under discussion. However, we wish to emphasise the *meaning* of integrals, rather than techniques for their evaluation. For practical purposes, their values can always be found to any required accuracy by direct computation of approximating sums (see the definitions P.1–P.5).

P.2 Surfaces and surface integrals

A closed surface S, or an open surface $S(\Gamma)$, can be dissected into elements, as is indicated in Fig. P.1. If the surface is smooth, and the elements are sufficiently small, each element can be approximated by an element of a plane. Two important properties can be defined for a plane surface element: the first is its area δS, and the second a unit vector $\hat{\mathbf{n}}$ that is normal to the plane. It is sometimes useful to consider the surface element as itself a vector $\delta\mathbf{S} = \hat{\mathbf{n}}\,\delta S$, of magnitude δS, pointing in the direction $\hat{\mathbf{n}}$.

The area of a surface is defined to be:

$$\text{area} = \int_S dS = \text{limit}\left(\sum_i \delta S_i\right), \tag{P.2}$$

where the linear dimensions of the δS_i go to zero in the limit.

In a similar way we can define integrals of functions over surfaces:

$$\int_S f(\mathbf{r})\,dS = \text{limit}\left(\sum_i f(\mathbf{r}_i)\,\delta S_i\right), \tag{P.3}$$

where $\mathbf{r}_i$ is a point in δS_i.

The case when $f(\mathbf{r}) = \mathbf{F}(\mathbf{r})\cdot\hat{\mathbf{n}}$ often occurs in physical theories, for example when we consider the flow of electric charge or energy across a surface. We may conveniently write

$$\int_S \mathbf{F}\cdot\hat{\mathbf{n}}\,dS = \int_S \mathbf{F}\cdot d\mathbf{S}.$$

It is important to make rules which specify the direction of the unit normal $\hat{\mathbf{n}}$.

For a surface S enclosing a volume V we always take the direction of $\hat{\mathbf{n}}$ to be pointing outwards from the volume (Fig. P.1).

For a two-sided surface $S(\Gamma)$, bounded by a curve Γ, we relate the direction of $\hat{\mathbf{n}}$ to the sense in which we go round Γ by a 'right-hand' rule

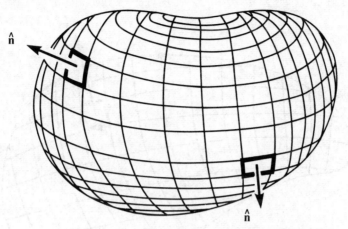

Fig. P.1 A surface S enclosing a volume V. The direction of the unit normal $\hat{\mathbf{n}}$ at a point on S is taken outwards from the interior of the volume. If the volume V surrounds interior cavities, S will consist of disjoint pieces.

explained in the caption to Fig. P.2. This rule only works if $S(\Gamma)$ is two-sided: we shall never consider one-sided surfaces like the famous 'Möbius strip'.

P.3 Line integrals

To define an integral along a line Γ in space joining points A and B, we dissect the line into elements which, for a sufficiently smooth curve, can be approximated by straight line elements of length δl pointing in the direction $\hat{\mathbf{t}}$, where $\hat{\mathbf{t}}$ is a unit vector tangent to the curve (Fig. P.3).

A vector line element $\delta\mathbf{l} = \hat{\mathbf{t}}\,\delta l$ can then be defined. The direction of $\hat{\mathbf{t}}$ is taken in the sense we choose to go along the curve Γ, from A to B say.

The total length of the curve is defined to be

$$\text{length} = \int_{\Gamma} dl = \text{limit}\left(\sum_{i} \delta l_i\right), \tag{P.4}$$

and we can define line integrals

$$\int_{\Gamma} f(\mathbf{r})\,dl = \text{limit}\left(\sum_{i} f(\mathbf{r}_i)\,\delta l_i\right), \tag{P.5}$$

where $\mathbf{r}_i$ is a point in the element δl_i.

Line integrals with $f(r) = F(\mathbf{r}) \cdot \hat{\mathbf{t}}$ often occur in physical theories, and we shall write

$$\int_{\Gamma} \mathbf{F} \cdot \hat{\mathbf{t}}\,dl = \int_{\Gamma} \mathbf{F} \cdot d\mathbf{l}.$$

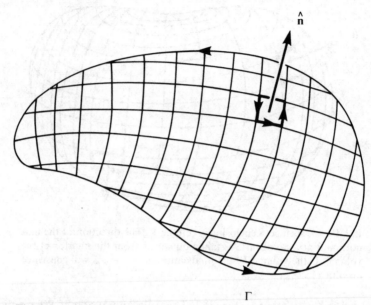

Fig. P.2 A curve Γ bounding a two-sided surface S. The surface has
been dissected into small pieces. The right-hand rule which relates the
sense of Γ to that of a unit normal $\hat{\mathbf{n}}$ to a surface element, is indicated.
At a point close to Γ, a right-handed screw through the surface,
turned in the sense of Γ, would be driven in the direction of $\hat{\mathbf{n}}$;
elsewhere the direction of $\hat{\mathbf{n}}$ is determined by continuity.

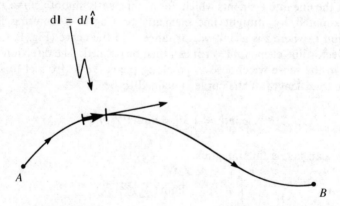

Fig. P.3 A directed line element $\delta\mathbf{l} = \hat{\mathbf{t}}\,\delta l$ of a curve Γ going from A
to B.

In going in the reverse direction from B to A, $\delta \mathbf{l} \rightarrow \delta \mathbf{l}' = -\delta \mathbf{l}$, and

$$\int_{\Gamma(AB)} \mathbf{F} \cdot d\mathbf{l} = -\int_{\Gamma(BA)} \mathbf{F} \cdot d\mathbf{l}. \tag{P.6}$$

If A and B coincide, Γ becomes a closed curve; we denote an integral round a closed curve by the symbol $\oint$.

P.4 The divergence theorem and Stokes's theorem

There are two important mathematical theorems which we use over and over again. The *divergence theorem* relates derivatives of a vector function $\mathbf{F}(\mathbf{r})$, throughout a volume V, to the values of that function on the surface enclosing the volume. The theorem states that

$$\blacktriangleright \qquad \int_V \nabla \cdot \mathbf{F} \, dV = \int_S \mathbf{F} \cdot \hat{\mathbf{n}} \, dS = \int_S \mathbf{F} \cdot d\mathbf{S}. \tag{P.7}$$

The direction of $\hat{\mathbf{n}}$ is given by the rule explained above.

Stokes's theorem relates derivatives of a vector function evaluated on *any* (two-sided) surface S bounded by the curve Γ, to values of that function on Γ, and states

$$\blacktriangleright \qquad \int_S (\nabla \times \mathbf{F}) \cdot d\mathbf{S} = \oint_\Gamma \mathbf{F} \cdot d\mathbf{l}. \tag{P.8}$$

The direction of $d\mathbf{S} = \hat{\mathbf{n}} \, dS$, and the sense of the integral round Γ, are related by the right-hand rule.

Along with these theorems there are the similar, and essentially equivalent, theorems:

$$\int_V \nabla \times \mathbf{F} \, dV = -\int_S \mathbf{F} \times d\mathbf{S}, \tag{P.9}$$

$$\int_V \nabla u \, dV = \int_S u \, d\mathbf{S}, \tag{P.10}$$

$$\int_S \nabla u \times d\mathbf{S} = -\oint_\Gamma u \, d\mathbf{l}. \tag{P.11}$$

We shall occasionally find these formulations useful.

The ability to relate physics inside a volume to physics on the enclosing surface, or physics on a surface to physics on the bounding curve, is of paramount importance in a field theory.

(Note that both the divergence theorem and Stokes's theorem can be thought of as simply extensions of the 'fundamental theorem of the integral calculus':

$$\int_a^b \frac{df}{dx}dx = f(b) - f(a),$$

which relates the derivative df/dx of a function $f(x)$ over an interval $[a, b]$ to the values of $f(x)$ at its end-points. This result is 'embedded' in three dimensions, or on a surface; proofs of the theorems along these lines will be found in standard mathematics texts.)

Problems

P.1   Obtain (P.9) and (P.10) from the divergence theorem (P.7) by taking (i) $\mathbf{F} = \mathbf{A} \times \mathbf{G}$, (ii) $\mathbf{F} = \mathbf{A}u$ in (P.7), where $\mathbf{A}$ is an arbitrary constant vector.

P.2   Obtain (P.11) from Stokes's theorem (P.8) by taking $\mathbf{F} = \mathbf{A}u$ in (P.8), where $\mathbf{A}$ is an arbitrary constant vector.

1

Charges and currents

1.1 Electric charge and conservation of charge

The basic constituents of matter, electrons and atomic nuclei, are all endowed with electric charge. It is through the electromagnetic fields generated by these charges that electrons and nuclei interact to form atoms and molecules and, hence, all materials. An electron carries a negative charge $-e$, and an atomic nucleus a positive charge Ze, where Z is an integer ranging from $Z = 1$ for hydrogen to $Z = 92$ for uranium (and higher for some unstable nuclei). The SI unit of charge is the *coulomb* (C) and $e \approx 1.602 \times 10^{-19}$ C.

The assignation of negative and positive sign is no more than a convention, which was set in the eighteenth century by the American physicist and statesman Benjamin Franklin. It is, however, a profound law of nature that, in an isolated system, the *net* total charge will never change: charge is conserved. In much of physics and chemistry neither an electron nor an atomic nucleus is ever created or destroyed, and charge conservation follows from this. More generally, the processes of nuclear physics do create and annihilate electrons, and transmute nuclei, but no known physical process can change the net total charge of an isolated system.

1.2 Charge density

At the present limits of experimental resolution, electrons seem to be 'point particles', in the sense that no intrinsic size or structure has yet been discerned for them. Treating the electron classically, we shall therefore define the *electron charge density* $\rho_{el}(\mathbf{r})$ of a single electron at the point $\mathbf{r}_1$ to be

$$\rho_{el}(\mathbf{r}) = -e\delta(\mathbf{r} - \mathbf{r}_1),$$

7

where the function $\delta(\mathbf{r} - \mathbf{r}_1)$ is defined to have the properties:

$$\delta(\mathbf{r} - \mathbf{r}_1) = 0, \text{ except when } \mathbf{r} = \mathbf{r}_1; \tag{1.1}$$

$$\int_V \delta(\mathbf{r} - \mathbf{r}_1) \, dV = 1, \tag{1.2}$$

when we integrate over any volume V containing the point $\mathbf{r}_1$.

Dirac introduced this useful function into mathematics, and it is known as the 'Dirac δ-function'; we have the three-dimensional version here. An important property which follows immediately from (1.1) and (1.2) is that, for any function $f(\mathbf{r})$,

$$\int_V f(\mathbf{r}) \, \delta(\mathbf{r} - \mathbf{r}_1) \, dV = f(\mathbf{r}_1), \tag{1.3}$$

if $\mathbf{r}_1$ lies inside V.

We can think of $\delta(\mathbf{r})$ as represented by (for example)

$$\delta(\mathbf{r}) = \frac{1}{\pi^{\frac{3}{2}} a^3} e^{-r^2/a^2}, \tag{1.4}$$

when the distance a becomes very small (Fig. 1.1 and Problem 1.1).

For the charge density of an assembly of electrons which are at the points $\mathbf{r}_i$, we have

$$\rho_{\text{el}}(\mathbf{r}) = \sum_i (-e) \, \delta(\mathbf{r} - \mathbf{r}_i). \tag{1.5}$$

On integrating (1.5) over a volume V, each electron inside V gives a contribution $(-e)$, and those outside V make no contribution, so that the total charge of the electrons contained in any volume V is given by:

$$\text{electronic charge in } V = \int_V \rho_{\text{el}}(\mathbf{r}) \, dV. \tag{1.6}$$

Atomic nuclei are found experimentally to have a structure on a scale of 10^{-15} m, and nuclear physicists define nuclear charge densities on this scale. On an atomic scale, where distances are measured in ångströms (1 Å $= 10^{-10}$ m), it is usually sufficient to regard the atomic nuclei as structureless point particles, and we can then define an *atomic charge density*

$$\rho_{\text{at}}(\mathbf{r}) = \sum_i Q_i \, \delta(\mathbf{r} - \mathbf{r}_i), \tag{1.7}$$

where the sum now is over all the particles (electrons and nuclei) in the system and Q_i is the charge on the particle at $\mathbf{r}_i$.

Equations (1.5) and (1.7) correspond to a classical picture. In quantum physics the charge densities are the expectation values of these expressions with respect to the wave functions of the electrons and nuclei; we obtain

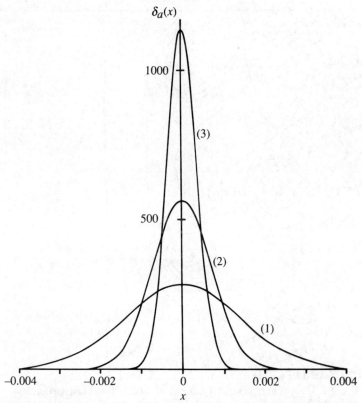

Fig. 1.1 $\delta(\mathbf{r}) = \delta(x)\,\delta(y)\,\delta(x)$, where $\delta(x)$ may be represented by $\delta_a(x) = e^{-x^2/a^2}/\pi^{\frac{1}{2}}a$, as $a \to 0$ (Problem 1.1). The figure shows plots of $\delta_a(x)$ for (1) $a = 0.002$, (2) $a = 0.001$, (3) $a = 0.0005$.

an averaged more smooth $\rho_{at}(\mathbf{r})$, but this still varies rapidly over atomic dimensions (Fig. 1.2). However, in a 'macroscopic' description of condensed matter, structure on a scale of less than, say, one micron (1 μm $= 10^{-6}$ m) is not considered. Then the charge densities $\rho(\mathbf{r})$ that are of interest are averages of the atomic charge densities over regions with dimensions L, where 1 Å $\ll L < 1$ μm. Mathematically, we can write this averaging as

$$\rho(\mathbf{r}) = \langle \rho_{at} \rangle = \int \rho_{at}(\mathbf{r}-\mathbf{r}')\,F(r')\,\mathrm{d}V', \qquad (1.8)$$

where $F(r)$ is some smooth isotropic weighting function of the shape indicated in Fig. 1.3, extending over a distance $\sim L$ and normalised so that

$$\int_V F(r')\,\mathrm{d}V' = 1. \qquad (1.9)$$

Fig. 1.2 Quantum mechanical smoothing of electronic charge density near the surface of copper: calculated contours of constant charge density in a plane normal to the (111) surface. The copper nuclei (solid dots) are treated as point charges in the calculation. The contour on the right is 10^{-3} in atomic units (ea_0^{-3}), and subsequent contours increase by factors of 10. (After Appelbaum, J. A. and Hamann, D. R. (1978), *Solid State Comm.*, **27**, 881.)

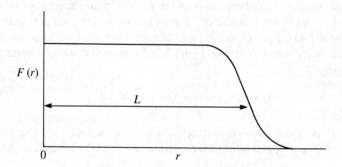

Fig. 1.3 A form for the smoothing function $F(r)$.

The resulting 'smeared out' $\rho(\mathbf{r})$ is a smooth function which can only change appreciably over distances $\gtrsim L$. Although $\rho(\mathbf{r})$ is not sensitive to the precise shape of $F(r)$, the appropriate value of L will depend on the nature of the system considered. If we take $L \sim 0.1$ μm in condensed matter, we shall be averaging over $\sim 10^9$ electrons and nuclei.

In this book we shall be principally concerned with establishing, and applying, the macroscopic equations of electromagnetism. It is quite possible to do this without mentioning the underlying atomic structures, but one then loses a great deal of insight into the concepts of the theory and, indeed, its limitations.

1.3 Current flow and the continuity equation

In many processes of physics and chemistry the outer, most weakly bound, of the atomic electrons can migrate from one atom to another. A neutral atom or molecule which loses one or more electrons is then positively charged and is called a positive ion ('cation'). A migrating electron can be captured by another neutral atom or molecule to form a negative ion ('anion'), or it may move more or less freely as a conduction electron, capable of transporting its charge through the system. In a solid it is usually only electrons which contribute to charge transport; in a liquid or a plasma, positive and negative ions may contribute to and may dominate in charge transport.

At the atomic level, we can define an *atomic current density*

$$\mathbf{J}_{\mathrm{at}}(\mathbf{r}, t) = \sum_i Q_i \mathbf{v}_i \delta(\mathbf{r} - \mathbf{r}_i(t)), \qquad (1.10)$$

where $\mathbf{v}_i = \mathrm{d}\mathbf{r}_i/\mathrm{d}t$ is the velocity of the ith particle. Whereas the charge density is a scalar field, the current density is a vector field. We have made explicit the dependence of $\mathbf{r}_i$ on t.

The charge density and current density are related by the continuity equation:

▶
$$\frac{\partial \rho_{\mathrm{at}}}{\partial t} + \mathbf{\nabla} \cdot \mathbf{J}_{\mathrm{at}} = 0. \qquad (1.11)$$

This follows directly from the definitions, since, for each particle i,

$$\frac{\partial}{\partial t} \delta(\mathbf{r} - \mathbf{r}_i(t)) = \frac{\partial \delta}{\partial x_i} \frac{\mathrm{d}x_i}{\mathrm{d}t} + \frac{\partial \delta}{\partial y_i} \frac{\mathrm{d}y_i}{\mathrm{d}t} + \frac{\partial \delta}{\partial z_i} \frac{\mathrm{d}z_i}{\mathrm{d}t}$$

$$= -\frac{\partial \delta}{\partial x} \frac{\mathrm{d}x_i}{\mathrm{d}t} - \frac{\partial \delta}{\partial y} \frac{\mathrm{d}y_i}{\mathrm{d}t} - \frac{\partial \delta}{\partial z} \frac{\mathrm{d}z_i}{\mathrm{d}t}$$

$$= -\mathbf{v}_i \cdot \mathbf{\nabla}\delta = -\mathbf{\nabla} \cdot (\mathbf{v}_i \delta).$$

(The second step holds since δ is a function of $(\mathbf{r} - \mathbf{r}_i)$, and the last since $\mathbf{r}_i$, and hence $\mathbf{v}_i = \mathrm{d}\mathbf{r}_i/\mathrm{d}t$, depends only on t.)

The physical interpretation of $\mathbf{J}_{at}$ is perhaps most easily seen by integrating (1.11) over a volume V bounded by an arbitrary fixed surface S. Then the $\partial/\partial t$ operator can be taken outside the integral, and the divergence theorem (P.7) can be used on the second term, to give

$$\frac{\partial}{\partial t}\int_V \rho_{at}\,\mathrm{d}V + \int_S \mathbf{J}_{at}\cdot\mathrm{d}\mathbf{S} = 0. \tag{1.12}$$

This is the integral form of the continuity equation. The first term is just the rate of change of the total charge inside S. We can identify $\mathbf{J}_{at}\cdot\mathrm{d}\mathbf{S}$ as the rate of flow of charge through the surface element $\mathrm{d}\mathbf{S}$. (This is also evident from contemplation of the definition (1.10)!) However, the continuity equation (1.11) expresses the law of charge conservation in its most simple and useful form.

1.4 Macroscopic current densities

The macroscopic current density $\mathbf{J}(\mathbf{r})$ at a point $\mathbf{r}$ is defined in a similar way to the macroscopic charge density: it is the atomic current density averaged over a volume (of the order of $(0.1\ \mu\mathrm{m})^3$ in condensed matter) around the point $\mathbf{r}$, with the weighting function $F(r)$ introduced in equation (1.8),

$$\mathbf{J}(\mathbf{r}) = \langle\mathbf{J}_{at}\rangle = \int \mathbf{J}_{at}(\mathbf{r} - \mathbf{r}')\,F(r')\,\mathrm{d}V'. \tag{1.13}$$

In materials this averaging involves $\sim 10^9$ particles and, although $\rho_{at}(\mathbf{r}, t)$ and $\mathbf{J}_{at}(\mathbf{r}, t)$ may vary rapidly with both time and position, macroscopic charge densities and currents can be quite slowly varying or even constant in both space and time.

Suppose, for simplicity, that only one kind of charged particle is responsible for the current; then the macroscopic current density is the macroscopic charge density ρ_c of those particles, multiplied by their mean velocity $\bar{\mathbf{v}}$ within the averaging volume:

▶ $$\mathbf{J}(\mathbf{r}) = \rho_c(\mathbf{r})\,\bar{\mathbf{v}}. \tag{1.14}$$

This follows immediately from the definitions (Problem 1.4).

The validity of the continuity equation on the atomic scale implies its validity on the macroscopic scale. Using (1.13),

$$\frac{\partial\rho}{\partial t} + \boldsymbol{\nabla}\cdot\mathbf{J} = \int \left[\frac{\partial\rho_{at}(\mathbf{r} - \mathbf{r}')}{\partial t} + \boldsymbol{\nabla}\cdot\mathbf{J}_{at}(\mathbf{r} - \mathbf{r}')\right] F(r')\,\mathrm{d}V',$$

(since both the $\partial/\partial t$ and the $\boldsymbol{\nabla}$ operators can be taken inside the integral,

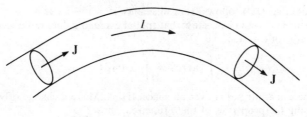

Fig. 1.4 In a wire carrying a steady current, $I = \int \mathbf{J} \cdot d\mathbf{S}$. The integral is over any cross-sectional area of the wire.

which is over $\mathbf{r}'$). But the quantity in square brackets vanishes at each point in space. Hence

▶
$$\frac{\partial \rho}{\partial t} + \nabla \cdot \mathbf{J} = 0. \tag{1.15}$$

This equation is of great importance in the development of our subject.

1.5 Steady currents in wires

You will be most familiar with the idea of electric currents flowing in wires. The conducting wire, basically a long thin cylinder of metal bent into some shape, was an early technological invention. Charge transport in wires is due to the flow of conduction electrons.

The electric current I through a cross-section of the wire is the charge per unit time passing through the cross-section (in a given sense) and is given by

$$I = \int_{\text{cross-section}} \mathbf{J} \cdot d\mathbf{S}. \tag{1.16}$$

In a steady state, and if electrons are not evaporating from the wire, there is no accumulation or loss of charge in the wire, and by the continuity equation I is constant along the wire (Fig. 1.4).

The SI unit of current is the *ampere* (A) and corresponds to a charge flow of one coulomb per second. Current densities have dimension A m^{-2}.

Problems

1.1 Show that $\delta(\mathbf{r}) = \delta(x)\,\delta(y)\,\delta(z)$, where $\delta(x)$ is the one-dimensional Dirac δ-function with the properties:

$\delta(x) = 0$, except when $x = 0$,

$\int \delta(x)\,dx = 1$, when the range of integration includes $x = 0$.

Show that $\delta(x)$ may be represented by

$$\delta_a(x) = e^{-x^2/a^2}/\pi^{\frac{1}{2}}a \text{ as } a \to 0.$$

Note that (1.4) follows from this, since $r^2 = x^2 + y^2 + z^2$.

1.2 Using (1.8) and (1.9), show that spatial averaging does not affect the total charge of a system, i.e.,

$$\int_V \rho(\mathbf{r})\,dV = \int_V \rho_{at}(\mathbf{r})\,dV,$$

where the integral is over all space. (Hint: Make a change of variables.)

1.3 Using the properties of the δ-function, show

$$\rho(\mathbf{r}) = \sum_i Q_i F(|\mathbf{r}-\mathbf{r}_i|).$$

$$\mathbf{J}(\mathbf{r}) = \sum_i Q_i \mathbf{v}_i F(|\mathbf{r}-\mathbf{r}_i|).$$

1.4 If there is only one type of charge carrier in a system, with charge density $\rho_c(\mathbf{r})$, use the results of Problem 1.3 to show

$$\mathbf{J}(\mathbf{r}) = \rho_c(\mathbf{r})\,\bar{\mathbf{v}}(\mathbf{r}),$$

where

$$\bar{\mathbf{v}}(\mathbf{r}) = \sum_i \mathbf{v}_i p(|\mathbf{r}-\mathbf{r}_i|) \text{ and } p(|\mathbf{r}-\mathbf{r}_i|) = \frac{F(|\mathbf{r}-\mathbf{r}_i|)}{\sum_i F(|\mathbf{r}-\mathbf{r}_i|)}.$$

Note that $\sum_i p(|\mathbf{r}-\mathbf{r}_i|) = 1$. The function $p(|\mathbf{r}-\mathbf{r}_i|)$ samples the velocities in the neighbourhood of $\mathbf{r}$.

1.5 The number density of conduction electrons in copper is 8.47×10^{28} m^{-3}. A copper wire of diameter 1 mm carries a current of 10 A. What is the mean velocity of the conduction electrons?

1.6 A cylindrical electron beam of the kind used in a television set has diameter 1 mm. The electrons have kinetic energy 10 keV and the total current is 100 μA. What is the electron velocity? What is the electron number density in the beam? In what time does an electron move the distance 0.5 m from the gun to the screen?

1.7 In the manufacture of aluminium, an electric current passes through a melt containing aluminium ions, mostly Al^{3+} which carry a charge of $3e$, and oxygen (O^{2-}) ions. The aluminium ions move to the negative electrode onto which they are deposited and the oxygen ions to the positive electrode. The melt remains electrically neutral. Estimate the mean current needed to produce one tonne (10^3 kg) of aluminium per day.

1.8 A 'toy model' of a crystal in one dimension consists of point ions of charge Q, lattice spacing a, in a uniform background of compensating electronic charge, so that

$$\rho_{at}(x) = Q\left[\sum_{n=-\infty}^{\infty} \delta(x-na) - (1/a)\right].$$

Here $\delta(x)$ is a one-dimensional δ-function.

Take the smoothing function $F(x) = (\pi L^2)^{-\frac{1}{2}} e^{-x^2/L^2}$ and show that

$$\rho(x) = \int_{-\infty}^{\infty} \rho_{at}(x-x')\,F(x')\,dx' = Q[\sum (\pi/L^2)^{-\frac{1}{2}} e^{-(x-na)^2/L^2} - (1/a)].$$

Write a computer program for $\rho(x)$ (including an appropriate number of terms) and show that even taking $L = 5a$, $\rho(x) \approx 0$ for all x.

2

Electrostatics

In the first chapter we stressed that electric charge is a conserved quantity. We now consider the forces between charges at rest: 'electrostatics', and introduce the concepts of the electrostatic field and the electrostatic potential.

2.1 Coulomb's law

The laws of electrostatics were deduced in the eighteenth century from experiments on macroscopic charged bodies. The force $\mathbf{F}_{21}$ on a small body located at position $\mathbf{r}_2$ and carrying charge Q_2, due to a small body at position $\mathbf{r}_1$ carrying charge Q_1, when the distance between them $\mathbf{r} = \mathbf{r}_2 - \mathbf{r}_1$ is large compared with their size, is given by

$$\mathbf{F}_{21} = (\text{constant})\frac{Q_1 Q_2}{r^2}\hat{\mathbf{r}}, \qquad (2.1)$$

where $\hat{\mathbf{r}}$ denotes the unit vector $\mathbf{r}/r$, and the constant of proportionality is positive. The constant depends on the unit of charge. (It also depends on whether the experiments are carried out in air or in a vacuum, but the difference is small, less than one part in a thousand.)

The natural philosophers of the eighteenth century were disposed to the idea of an 'inverse square' law for electrostatics, following the success of Newton's inverse square law for gravitation. However, it was Coulomb's careful experiments in 1785 using his very sensitive torsion balance, on the force between like charges, followed by further investigations in 1787 to cover the attractive case, which gave direct quantitative verification of the inverse square law; it is now universally known as *Coulomb's law*.

Coulomb used gilded pith balls of about 2 mm radius, with separations varied over distances of a few centimetres. Today there is much indirect evidence that his law holds on all length scales. For example, our almost complete understanding of the hydrogen atom exemplifies its truth on an

15

atomic scale, and its validity in electron–electron interactions has been probed in high energy scattering experiments down to distances of 10^{-17} m.

In the SI system of units the unit of charge is the *coulomb* (C), and in these units the force between two 'point' charges is given by

$$\blacktriangleright \qquad \mathbf{F}_{21} = \frac{1}{4\pi\varepsilon_0} \frac{Q_1 Q_2}{r^2} \hat{\mathbf{r}} = \frac{1}{4\pi\varepsilon_0} \frac{Q_1 Q_2 (\mathbf{r}_2 - \mathbf{r}_1)}{|\mathbf{r}_2 - \mathbf{r}_1|^3}, \qquad (2.2)$$

where $\varepsilon_0 \approx 8.854 \times 10^{-12}$ C^2 N^{-1} m^{-2} is called the *permittivity of free space*. The numerical value of ε_0 will be explained in Chapter 6.

We regard charged bodies as point charges when their dimensions are small compared with values of r of interest, and their internal structure may be neglected. For example (cf. Chapter 1), an atomic nucleus is a complex structure to a nuclear physicist but (usually) a point charge to an atomic or condensed matter physicist. In some situations it may suffice to treat ions as point particles.

2.2 The superposition principle and the electric field

To develop the theory of electrostatics based on Coulomb's law we need also another fundamental experimental law, the *superposition principle*: the total force on a point charge Q at $\mathbf{r}$ is the vector sum of the forces due to the other point charges Q_i in the system:

$$\mathbf{F}(\mathbf{r}) = \frac{1}{4\pi\varepsilon_0} \sum_i \frac{Q Q_i (\mathbf{r} - \mathbf{r}_i)}{|\mathbf{r} - \mathbf{r}_i|^3}. \qquad (2.3)$$

We can write this as

$$\mathbf{F}(\mathbf{r}) = Q\mathbf{E}(\mathbf{r}), \qquad (2.4)$$

where

$$\mathbf{E}(\mathbf{r}) = \frac{1}{4\pi\varepsilon_0} \sum_i \frac{Q_i (\mathbf{r} - \mathbf{r}_i)}{|\mathbf{r} - \mathbf{r}_i|^3} \qquad (2.5)$$

is called the *electric field* at $\mathbf{r}$ due to the charges Q_i.

Provided that the positions of the charges Q_i are not influenced by introducing Q, we have through equation (2.4) an 'operational definition' for the field $\mathbf{E}(\mathbf{r})$ at any point in space: $\mathbf{E}(\mathbf{r})$ may be ascertained by measuring the force required to hold a 'test body' carrying known charge Q at rest at the position $\mathbf{r}$.

2.3 The electrostatic potential

It is important to appreciate that the vector function $\mathbf{r}/r^3$ can be expressed as the gradient of a scalar function:

$$\frac{\mathbf{r}}{r^3} = \frac{\hat{\mathbf{r}}}{r^2} = -\nabla\left(\frac{1}{r}\right).$$

More generally, by a shift of origin to $\mathbf{r}_1$,

$$\frac{\mathbf{r} - \mathbf{r}_1}{|\mathbf{r} - \mathbf{r}_1|^3} = -\nabla\left(\frac{1}{|\mathbf{r} - \mathbf{r}_1|}\right). \tag{2.6}$$

Using this mathematical result, we can write (2.5) as

$$\mathbf{E}(\mathbf{r}) = -\nabla\Phi(\mathbf{r}), \tag{2.7}$$

where

$$\Phi(\mathbf{r}) = \frac{1}{4\pi\varepsilon_0} \sum_i \frac{Q_i}{|\mathbf{r} - \mathbf{r}_i|}. \tag{2.8}$$

The function $\Phi(\mathbf{r})$ is called the *electrostatic potential* due to the charges Q_i. Since it is a scalar field it is much simpler to handle than the vector field $\mathbf{E}(\mathbf{r})$. To the superposition principle corresponds the additivity of potentials.

The SI unit of potential is the *volt* (V). Electric fields are expressed in volts per metre (V m^{-1}).

The potential also has an important physical interpretation. From equation (2.4), the line integral

$$-\int_\Gamma \mathbf{E} \cdot \mathrm{d}\mathbf{l},$$

taken along a path Γ between points $\mathbf{r}_1$ and $\mathbf{r}_2$, represents the work done per unit positive charge *against* the electric field in moving a small charged test body along this path (Fig. 2.1), and

$$-\int_\Gamma \mathbf{E} \cdot \mathrm{d}\mathbf{l} = \int_\Gamma \nabla\Phi \cdot \mathrm{d}\mathbf{l} = \Phi(\mathbf{r}_2) - \Phi(\mathbf{r}_1) \tag{2.9}$$

(since $\nabla\Phi \cdot \mathrm{d}\mathbf{l} = (\partial\Phi/\partial x)\,\mathrm{d}x + (\partial\Phi/\partial y)\,\mathrm{d}y + (\partial\Phi/\partial z)\,\mathrm{d}z = \mathrm{d}\Phi$).

Thus the work done per unit charge does not depend on the particular path Γ, and is given by the difference in electrostatic potential between the two points.

From the definition (2.8), $\Phi \to 0$ at points far removed from the charge distribution. (Apart from occasional mathematical abstractions, we shall always consider systems of charges distributed over finite regions of space.) Hence from (2.9) we can interpret $\Phi(\mathbf{r})$ as the energy required to bring up a unit test charge from 'infinity' (where $\Phi \to 0$) to the point $\mathbf{r}$.

Since $\mathbf{E} = -\nabla\Phi$, and $\nabla \times \nabla\Phi \equiv 0$, the electrostatic field satisfies the condition

▶ $$\nabla \times \mathbf{E}(\mathbf{r}) = 0 \tag{2.10}$$

everywhere. Conversely, the condition $\nabla \times \mathbf{E} = 0$ implies, mathematically, the existence of some potential function such that $\mathbf{E} = -\nabla\Phi$. (However, this condition alone places no restriction on the form the potential might take; it does not imply an inverse square law of force.)

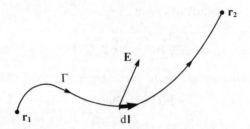

Fig. 2.1 $\mathbf{F} \cdot \mathbf{dl} = Q\mathbf{E} \cdot \mathbf{dl}$ is the work done on a particle carrying charge Q by the electric field in the displacement $\mathbf{dl}$ along the curve Γ.

2.4 Spatial averaging

The expressions we have obtained are believed to be valid at the atomic level, and we can write

$$\Phi_{\text{at}}(\mathbf{r}) = \frac{1}{4\pi\varepsilon_0} \int \frac{\rho_{\text{at}}(\mathbf{r}')}{|\mathbf{r}-\mathbf{r}'|} \, dV', \qquad (2.11)$$

where the integral is taken over all space; ρ_{at} is given by (1.7):

$$\rho_{\text{at}}(\mathbf{r}') = \sum_i Q_i \, \delta(\mathbf{r}'-\mathbf{r}_i).$$

Using the property (1.3) of the δ-function, (2.11) is equivalent to (2.8). In a quantum picture, $\rho_{\text{at}}(\mathbf{r}')$ becomes the smooth expectation value of the charge density (Fig. 1.2). (We shall often, throughout the book, denote by $\mathbf{r}$ the point at which we evaluate a field quantity, and by $\mathbf{r}'$ a point of a distribution over which we integrate.)

Applying the spatial averaging procedure of §1.1 to (2.11) gives

$$\Phi(\mathbf{r}) = \langle \Phi_{\text{at}} \rangle = \int \Phi_{\text{at}}(\mathbf{r}-\mathbf{r}'') F(r'') \, dV''$$

$$= \frac{1}{4\pi\varepsilon_0} \iint \frac{\rho_{\text{at}}(\mathbf{r}')}{|\mathbf{r}-\mathbf{r}''-\mathbf{r}'|} F(r'') \, dV' \, dV''.$$

Making the change of variables $\mathbf{r}_1 = \mathbf{r}'+\mathbf{r}''$, $\mathbf{r}_2 = \mathbf{r}''$, we have $dV' \, dV'' = dV_1 \, dV_2$ (since the Jacobian of the transformation is unity) so that

$$\Phi(\mathbf{r}) = \frac{1}{4\pi\varepsilon_0} \iint \frac{\rho_{\text{at}}(\mathbf{r}_1-\mathbf{r}_2)}{|\mathbf{r}-\mathbf{r}_1|} F(r_2) \, dV_1 \, dV_2.$$

The integration over V_2 then yields

$$\blacktriangleright \qquad \Phi(\mathbf{r}) = \frac{1}{4\pi\varepsilon_0} \int \frac{\rho(\mathbf{r}_1)}{|\mathbf{r}-\mathbf{r}_1|} \, dV_1, \qquad (2.12)$$

where $\rho(\mathbf{r}_1)$ is the averaged charge density. This is the expression you might have anticipated!

The macroscopic electric field is given by

$$\mathbf{E}(\mathbf{r}) = \langle -\nabla\Phi_{at}\rangle = -\nabla\langle\Phi_{at}\rangle.$$

Hence at all points

$$\blacktriangleright \qquad \mathbf{E} = -\nabla\Phi, \qquad (2.13)$$

and $\mathbf{E}(\mathbf{r}_1)$ is therefore perpendicular to that *equipotential surface*, $\Phi(\mathbf{r}) =$ constant, which passes through $\mathbf{r}_1$.

In contrast to the underlying atomic expressions from which they are derived, the expressions (2.12) and (2.13) for the averaged fields $\Phi(\mathbf{r})$ and $\mathbf{E}(\mathbf{r})$ remain finite even at points $\mathbf{r}$ lying inside the charge distribution, and are everywhere continuous functions of $\mathbf{r}$.

The experiments described in §2.1 and §2.2 measure averaged fields, so that the notation is consistent with its use in these earlier paragraphs!

2.5 Gauss's theorem and the field equations

The *electric flux* through an (imagined) surface S is defined to be

$$\text{flux} = \int_S \mathbf{E}\cdot d\mathbf{S}.$$

Gauss's theorem relates the flux through a *closed* surface S to the charge enclosed by that surface. The basic mathematical result, unique to and characteristic of the inverse square law, is that, given a point charge Q at $\mathbf{R}$ producing a field $\mathbf{E}$,

$$\int_S \mathbf{E}\cdot d\mathbf{S} = \begin{cases} Q/\varepsilon_0 & \text{if } \mathbf{R} \text{ lies inside } S, \\ 0 & \text{if } \mathbf{R} \text{ lies outside } S. \end{cases} \qquad (2.14)$$

You have probably met this theorem before; we give a proof in Appendix A.

It follows using the superposition principle that

$$\int_S \mathbf{E}_{at}\cdot d\mathbf{S} = \frac{1}{\varepsilon_0}\int_V \rho_{at}\, dV. \qquad (2.15)$$

This is *Gauss's theorem* for the atomic field. It may be used directly to find the electric field of atomic charge distributions of high symmetry (Problems 2.2, 2.3, 2.5).

Using the divergence theorem (P.7) on the left-hand side, we can write (2.15) as

$$\int_V \nabla\cdot\mathbf{E}_{at}\, dV = \frac{1}{\varepsilon_0}\int_V \rho_{at}\, dV.$$

Since this holds for *any* volume V, the integrands on each side must be the same, so that

$$\nabla \cdot \mathbf{E}_{at}(\mathbf{r}) = \rho_{at}(\mathbf{r})/\varepsilon_0 \qquad (2.16)$$

at every point in space.

A similar equation holds on the macroscopic length scale; spatial averaging gives

▶ $$\nabla \cdot \mathbf{E}(\mathbf{r}) = \rho(\mathbf{r})/\varepsilon_0. \qquad (2.17)$$

This is a *field equation*, relating derivatives of $\mathbf{E}$ at a point to the charge density $\rho(\mathbf{r})$ at that point. Integrating (2.17) over a volume V gives the macroscopic form of Gauss's theorem:

▶ $$\int_S \mathbf{E} \cdot d\mathbf{S} = \frac{1}{\varepsilon_0} \int_V \rho \, dV. \qquad (2.18)$$

Since $\mathbf{E} = -\nabla \Phi$, the potential satisfies *Poisson's equation*:

▶ $$\nabla^2 \Phi(\mathbf{r}) = -\rho(\mathbf{r})/\varepsilon_0. \qquad (2.19)$$

Our formula (2.12) for the potential (replacing $\mathbf{r}_1$ by $\mathbf{r}'$),

$$\Phi(\mathbf{r}) = \frac{1}{4\pi\varepsilon_0} \int \frac{\rho(\mathbf{r}')}{|\mathbf{r}-\mathbf{r}'|} dV',$$

may be regarded as the solution of the differential equation (2.19). Mathematically, we could add to this any solution of the 'homogeneous' equation $\nabla^2 \Phi = 0$; for example, adding a potential which represents a uniform background electric field $(0, 0, E_0)$ gives

$$\Phi(\mathbf{r}) = \frac{1}{4\pi\varepsilon_0} \int \frac{\rho(\mathbf{r}')}{|\mathbf{r}-\mathbf{r}'|} dV' + A - E_0 z, \qquad (2.20)$$

where A is a constant, but additional terms like these are not zero at large distances and represent contributions to the potential from 'outside' charges. The expression (2.12) is the unique solution for the given charge distribution $\rho(\mathbf{r})$.

The free-space version of (2.19),

▶ $$\nabla^2 \Phi = 0 \qquad (2.21)$$

is known as *Laplace's equation*. It is useful in determining the potential in some situations when the charge distribution is not known a priori, and we shall return to it in Chapter 7.

The two field equations

$$\nabla \times \mathbf{E} = 0, \quad \nabla \cdot \mathbf{E} = \rho/\varepsilon_0, \qquad (2.22)$$

encapsulate the properties of the electrostatic field $\mathbf{E}(\mathbf{r})$. The first field equation ensures the existence of a potential function. The second field equation incorporates the superposition principle, and leads to Poisson's equation and its unique solution (2.12) for the potential. From the potential for a point charge we recover Coulomb's law.

2.6 Considerations on the energy of charge distributions

The potential due to a particle carrying a charge of Q_1 at $\mathbf{r}_1$ is

$$\Phi_1(\mathbf{r}) = \frac{1}{4\pi\varepsilon_0} \frac{Q_1}{|\mathbf{r} - \mathbf{r}_1|}.$$

Hence the energy required to bring another particle, charge Q_2, from infinity to $\mathbf{r}_2$, in the presence of Q_1, is

$$U_{12} = \frac{1}{4\pi\varepsilon_0} \frac{Q_1 Q_2}{|\mathbf{r}_2 - \mathbf{r}_1|}.$$

If a third particle, charge Q_3, is then brought up to $\mathbf{r}_3$, the total energy is

$$U_{123} = \frac{1}{4\pi\varepsilon_0} \left(\frac{Q_1 Q_2}{|\mathbf{r}_2 - \mathbf{r}_1|} + \frac{Q_1 Q_3}{|\mathbf{r}_3 - \mathbf{r}_1|} + \frac{Q_2 Q_3}{|\mathbf{r}_3 - \mathbf{r}_2|} \right).$$

Bringing charged particles sequentially from infinity, we can see that the electrostatic energy of an assembly of particles is in general

$$U = \frac{1}{2} \frac{1}{4\pi\varepsilon_0} \sum_{i,j}' \frac{Q_i Q_j}{|\mathbf{r}_i - \mathbf{r}_j|}, \tag{2.23}$$

where the prime denotes that terms $i = j$ are omitted from the sum, and the factor $(\frac{1}{2})$ arises since the sum counts each term in U twice.

   This expression can be written in another, instructive, form. We first add to U the terms with $i = j$. Each additional term represents the 'self energy' of a charge with itself, and, for truly point charges, is infinite. We shall ignore this difficulty, and write, using (1.7) and (2.11)

$$\text{energy} = \frac{1}{2} \sum_i Q_i \sum_j \frac{Q_j}{4\pi\varepsilon_0 |\mathbf{r}_i - \mathbf{r}_j|}$$

$$= \frac{1}{2} \frac{1}{4\pi\varepsilon_0} \iint \frac{\rho_{\mathrm{at}}(\mathbf{r}) \rho_{\mathrm{at}}(\mathbf{r}')}{|\mathbf{r} - \mathbf{r}'|} \, \mathrm{d}V \mathrm{d}V' \tag{2.24}$$

$$= \frac{1}{2} \int_V \rho_{\mathrm{at}}(\mathbf{r}) \Phi_{\mathrm{at}}(\mathbf{r}) \, \mathrm{d}V. \tag{2.25}$$

In (2.25), the integration can be over any volume V which includes all the charges. We shall take the volume to be a sphere of sufficiently large radius R to enclose the system, and consider the limit $R \to \infty$.

Since $\nabla \cdot \mathbf{E}_{at} = \rho_{at}/\varepsilon_0$, we can write

$$\text{energy} = \frac{\varepsilon_0}{2} \int \Phi_{at} \, \nabla \cdot \mathbf{E}_{at} \, dV$$

$$= -\frac{\varepsilon_0}{2} \int \mathbf{E}_{at} \cdot \nabla \Phi_{at} \, dV + \frac{\varepsilon_0}{2} \int \nabla \cdot (\mathbf{E}_{at} \, \Phi_{at}) \, dV$$

(using the formula $\nabla \cdot (\Phi \mathbf{E}) = \Phi \nabla \cdot \mathbf{E} + \mathbf{E} \cdot \nabla \Phi$).

With the help of the divergence theorem the last term can be transformed into the surface integral $(\varepsilon_0/2) \int \Phi_{at} \, \mathbf{E}_{at} \cdot d\mathbf{S}$, over the sphere of radius R. For a distribution of charge of finite extent it is clear from (2.5) and (2.8) that $\mathbf{E}_{at} \sim 1/R^2$ and $\Phi_{at} \sim 1/R$ as $R \to \infty$. The surface area of a sphere is $4\pi R^2$, so the surface integral $\to 0$ as $R \to \infty$. Hence, since $\mathbf{E}_{at} = -\nabla \Phi_{at}$,

$$\text{energy} = \frac{\varepsilon_0}{2} \int \mathbf{E}_{at}^2 \, dV, \tag{2.26}$$

where the integral is now over all space.

The formula (2.26) suggests that the energy of the system resides in the electric field, with an energy density $(\varepsilon_0/2) \mathbf{E}_{at}^2$; this interpretation will be most significant when we come to consider electromagnetic radiation. The self energies of the particles are included in (2.26), but usually it is only energy changes that are of interest and the self energies do not contribute to these.

Equations like (2.16) which are linear in the fields have the same form on the atomic scale as, after smoothing, on the macroscopic scale. This is *not* the case for the expressions (2.24), (2.25), and (2.26) for the energy, which are quadratic functions of the fields. We have established these expressions at the atomic level (and on this scale the energy density includes, for example, the electrostatic contributions to the chemical binding energies). Expressions for energies associated with *macroscopic* electric fields in materials are obtained in Chapter 10. In free space, away from materials, static electric fields do not change appreciably on micron length scales and the space averaged field $\mathbf{E}(\mathbf{r}) = \mathbf{E}_{at}(\mathbf{r})$; to a good approximation the electric energy density *in free space* $= (\varepsilon_0/2) |\mathbf{E}(\mathbf{r})|^2$.

We have ignored the difficulty that the self energy of a truly point particle seems to be infinite. In general, the self energy of a particle of mass m is given by the Einstein relation, $E = mc^2$. If the particle is charged, the electric self energy (divided by c^2) is just one contribution to its mass, so that in including electric self energies in equation (2.24) we are only including a part of the self energies. Even in the most successful theories of matter, such as quantum electrodynamics, there remain difficulties associated with point particle self energies.

Problems

2.1   Show that an electric field $\mathbf{E}_0$ which is uniform in some region of space is given, in that region, by the potential

$$\Phi(\mathbf{r}) = \Phi(0) - \mathbf{E}_0 \cdot \mathbf{r}.$$

2.2   A uniform spherical shell of charge is of radius R, and has total charge Q. Use Gauss's theorem to show that inside the shell the electric field is zero, and outside the field is the same as if the total charge Q were at the centre of the shell.

2.3   Use Gauss's theorem to obtain expressions for the electric field at points inside and outside a uniform spherical distribution of charge, of total charge Q.

        Hence find, by integration, the corresponding potential functions.

2.4   The electron charge density in a hydrogen atom in its ground state is $\rho_{el}(r) = (-e/\pi a_0^3)\exp(-2r/a_0)$, where r is the distance from the proton and a_0 is the Bohr radius.

        By solving Poisson's equation, show that the *total* electrostatic potential

$$\Phi_{at}(r) = \frac{e}{4\pi\varepsilon_0}\left(\frac{1}{r}+\frac{1}{a_0}\right)\exp\left(-2r/a_0\right).$$

        (Note that $\Phi_{at}(r)$ must behave correctly near $r = 0$, and as $r \to \infty$.)

2.5   Use Gauss's theorem to show that the electric field due to a uniform distribution of charge on a thin straight wire, of length l and total charge Q, at points away from the ends of the wire and distance ρ from the wire, where $\rho \ll l$, is radial from the wire and of strength

$$E(\rho) = \frac{(Q/l)}{2\pi\varepsilon_0\rho}.$$

        What is the corresponding potential function?

2.6   The direction of the electron beam of Problem 1.6 is changed by passing the beam through a region of electric field transverse to the initial beam direction. Estimate the magnitude of the electric field needed to turn the beam through $10°$, if the field is uniform and extends over a distance of 4 cm.

2.7   Initially the beam of Problem 1.6 is collimated and of uniform density. Estimate the radial electric field at the fringe of the beam, and hence estimate the radial spread of the beam in traversing the tube.

2.8   Show that the mean value of the electric potential inside a uniformly distributed spherical distribution of charge is

$$\bar{\Phi} = \frac{6}{5}\frac{Q}{4\pi\varepsilon_0 R}$$

where Q is the total charge, and R is the radius of the sphere.

        An atomic nucleus contains N neutrons and Z protons. Consider building up the nucleus sequentially, adding one nucleon at a time, and supposing the nucleons become distributed uniformly over a sphere of radius R. Show that the electrostatic contribution to the energy of the system is

$$\text{electrostatic energy} = \frac{3}{5}\frac{Z(Z-1)e^2}{4\pi\varepsilon_0 R}.$$

What is this energy for a uranium $(Z = 92)$ nucleus, taking $R = 6.8 \times 10^{-15}$ m? Express your answer in MeV.

2.9   Suppose an electron at rest is not a point object, but a spherical shell of charge of radius a. Find a, on the assumption that all the mass of the electron is electrostatic in origin, so that the field energy may be equated to $m_e c^2$.

2.10   An *electric field line* is a line whose tangent at any point on it gives the direction of the electric field at that point.

Show that the field lines corresponding to a potential field $\Phi(\mathbf{r})$ are given by the solutions of the differential equations

$$\frac{\mathrm{d}x}{(\partial\Phi/\partial x)} = \frac{\mathrm{d}y}{(\partial\Phi/\partial y)} = \frac{\mathrm{d}z}{(\partial\Phi/\partial z)}.$$

2.11   A long cylinder of radius R lies along the z-axis perpendicular to a uniform electric field $(E_0, 0, 0)$. The surface of the cylinder is uniformly charged with charge per unit length $-4\pi\varepsilon_0 E_0 R$. Show that the potential outside the cylinder (neglecting end effects) is (cf. Problem 2.5)

$$\Phi(x, y) = E_0[-x + 2R\ln(\sqrt{(x^2 + y^2)}/R)].$$

Show there is a *null point* $(\mathbf{E} = 0)$ at $(2R, 0)$.

Write a program to plot electric field lines within a square of side $10R$ with the cylinder at the centre.

In a gas at an appropriate pressure, charged particles would drift along the field lines. Estimate the proportion of positive ions produced by ionisation with uniform probability within the square that will migrate to the cylinder.

2.12   A particle detector consists of a fine wire of radius 10 μm on the axis of a long cylinder, radius 1 cm. The cylinder is held at zero potential and the wire at 3.5 kV. The device is gas filled with atoms that are ionised if struck by a free electron with kinetic energy greater than 4 eV. The distance between collisions for free electrons in the gas is 0.1 μm. Assume that such a collision results in two free electrons, each with zero kinetic energy.

Estimate the distance from the wire at which a free electron will cause ionisation, and the number of electrons arriving at the wire initiated by one electron in the detector.

3

Electric dipoles

In this short chapter we derive useful approximate expressions for the potential of a distribution of charge with density $\rho(\mathbf{r})$, localised in some region of space, when the potential is evaluated at points far from the distribution.

3.1 The electric dipole moment

The potential of a charge distribution is given by (2.12):

$$\Phi(\mathbf{r}) = \frac{1}{4\pi\varepsilon_0} \int \frac{\rho(\mathbf{r}')}{|\mathbf{r} - \mathbf{r}'|} \, dV'$$

$$= \frac{1}{4\pi\varepsilon_0} \int \frac{\rho(\mathbf{r}')}{(r^2 + r'^2 - 2\mathbf{r}\cdot\mathbf{r}')^{\frac{1}{2}}} \, dV',$$

since $|\mathbf{r} - \mathbf{r}'|^2 = (\mathbf{r} - \mathbf{r}')\cdot(\mathbf{r} - \mathbf{r}') = r^2 + r'^2 - 2\mathbf{r}\cdot\mathbf{r}'$.

We choose the origin of coordinates to lie within the distribution. At points $\mathbf{r}$ which are distant from the distribution, $r \gg r'$, so that $(r'^2 - 2\mathbf{r}\cdot\mathbf{r}')/r^2$ is small. Using the binomial expansion

$$(1 + x)^{-\frac{1}{2}} = 1 - \tfrac{1}{2}x + \tfrac{3}{8}x^2 \ldots,$$

we obtain to first order in r'/r:

$$\frac{1}{|\mathbf{r} - \mathbf{r}'|} = \frac{1}{r} + \frac{\mathbf{r}\cdot\mathbf{r}'}{r^3}.$$

Hence at distant points r,

$$\Phi(\mathbf{r}) \approx \frac{1}{4\pi\varepsilon_0}\left(\frac{Q}{r} + \frac{\mathbf{p}\cdot\mathbf{r}}{r^3}\right), \tag{3.1}$$

where

$$Q = \int \rho(r') \, dV'$$

is the net charge of the distribution, and

▶
$$\mathbf{p} = \int \rho(\mathbf{r'}) \, \mathbf{r'} \, dV' \tag{3.2}$$

is called the *electric dipole moment* of the distribution.

If the total charge Q is not zero, the potential at large distances is approximately that of a point charge at the origin, and the dipole term represents a correction. However, for a system of charges which is, overall, neutral, the dipole potential takes over as leading term. In this case, the dipole moment does not depend on the choice of origin (Problem 3.1); for example, point charges Q_1 at $(\mathbf{r}_1 + \mathbf{d})$ and $-Q_1$ at $\mathbf{r}_1$ have a dipole moment

$$\mathbf{p} = Q_1(\mathbf{r}_1 + \mathbf{d}) - Q_1 \mathbf{r}_1 = Q_1 \mathbf{d}; \tag{3.3}$$

this is in the direction of the displacement $\mathbf{d}$ of the positive charge relative to the negative charge.

On the atomic scale, a neutral atom in isolation has its negative electron cloud symmetrically distributed around its positive nucleus, and it has no discernible dipole moment. However, an external electric field will pull the electron cloud and nucleus in opposite directions, and thereby induce an electric dipole moment.

For example, the dipole moment of a hydrogen atom in an applied static electric field $\mathbf{E}$ is

$$\mathbf{p} = 4\pi\varepsilon_0 \alpha \mathbf{E}, \tag{3.4}$$

where $\alpha = 9a_0^3/2$, and $a_0 = 4\pi\varepsilon_0 \hbar^2/m_e e^2 = 0.529$ Å is the Bohr radius. Thus the dipole moment is proportional to the applied electric field. The coefficient $4\pi\varepsilon_0 \alpha$ is called the *atomic polarisability*.

The result (3.4) follows from a careful quantum mechanical calculation, but it can be qualitatively understood from a model in which the electron charge is distributed uniformly inside a sphere of radius a. The electric field of such a charge distribution at position $\mathbf{r}(r < a)$ relative to the centre is $-e\mathbf{r}/4\pi\varepsilon_0 a^3$ (Problem 2.3). In an external field $\mathbf{E}$, the atomic nucleus will be in equilibrium when the net force acting on it is zero. It is, then, displaced from the electronic centre of charge by $\mathbf{r}_n$ where

$$e\mathbf{E} - e^2\mathbf{r}_n/4\pi\varepsilon_0 a^3 = 0,$$

or

$$e\mathbf{r}_n = 4\pi\varepsilon_0 a^3 \mathbf{E}.$$

But the dipole moment is $\mathbf{p} = e\mathbf{r}_n$, so in this model we obtain

$$\mathbf{p} = 4\pi\varepsilon_0 a^3 \mathbf{E}.$$

This is similar in magnitude to (3.4) if $a \sim 1$ Å. In fact atomic polarisabilities are all found experimentally to be of magnitude $\sim 4\pi\varepsilon_0 \times$ (atomic volume).

Many molecules possess a permanent electric dipole moment associated

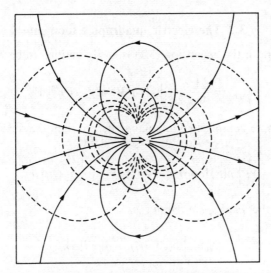

Fig. 3.1 Some electric field lines (with arrows) and sections of equipotential surfaces (dotted lines), for a dipole field. The dipole lies in the plane of the paper, as indicated.

with the orientation of the molecules, existing independently of any external electric field. For example, in the formation of a molecule of hydrogen chloride, HCl, the electronic charge distribution arising from the hydrogen atom is shifted towards the chlorine atom. If $\mathbf{a}$ is the position of the hydrogen nucleus relative to the chlorine nucleus and $\mathbf{p}$ the dipole moment of the molecule, the measured values of $|\mathbf{a}|$ and $|\mathbf{p}|$ (1.27 Å and 3.6×10^{-30} C m respectively), give $\mathbf{p} = 0.18e\mathbf{a}$. Comparing with equation (3.3) we see that the chemical combination of the two atoms results in a significant movement of charge.

The electric field associated with a dipole moment $\mathbf{p}$ is readily calculated from the dipole potential Φ_d (Problem 3.2). For a dipole at the origin,

$$\Phi_d = \frac{\mathbf{p} \cdot \mathbf{r}}{4\pi\varepsilon_0 r^3},$$

$$\mathbf{E} = -\nabla\Phi_d = \frac{1}{4\pi\varepsilon_0 r^3}\left(\frac{3(\mathbf{p} \cdot \mathbf{r})\mathbf{r}}{r^2} - \mathbf{p}\right). \tag{3.5}$$

Note that the field falls off with distance as r^{-3}. Some of the features of the dipole field pattern are illustrated in Fig. 3.1. The potential and field have axial symmetry about the direction of the dipole.

3.2 The electric quadrupole moment

The next term in the expansion (3.1) is easily verified to be

$$\Phi_{\text{quad}} = \frac{1}{4\pi\varepsilon_0}\frac{1}{2r^5}\int\rho(\mathbf{r}')[3(\mathbf{r}\cdot\mathbf{r}')^2 - r^2 r'^2]\,dV'. \tag{3.6}$$

This has a more complicated angular structure than the dipole term. We can write

$$\frac{1}{2}\int\rho(\mathbf{r}')[3(\mathbf{r}\cdot\mathbf{r}')^2 - r^2 r'^2]\,dV' = \frac{1}{2}\sum_{i,j} Q_{ij} r_i r_j,$$

where Q_{ij} is the *quadrupole tensor*:

$$Q_{ij} = \int\rho(\mathbf{r}')[3r_i' r_j' - \delta_{ij} r'^2]\,dV'. \tag{3.7}$$

Here we are using the notation $\mathbf{r} = (r_1, r_2, r_3)$ and the indices i, j run from 1 to 3. δ_{ij} is the usual 'Kronecker δ'.

In general a molecule can have a permanent electric quadrupole moment. In some cases such moments are significant in intermolecular interactions.

Further terms in the 'multipole expansion' can be systematically calculated, but those we have considered suffice for most purposes.

3.3 The force and torque on a charge distribution in an external electric field

We shall consider first the case of a charge distribution so confined in space that the external field may be considered a constant field, $\mathbf{E}_0$, throughout the distribution. Then the net force acting is the sum of the forces on the individual parts of the system:

$$\text{Force} = \int\rho(\mathbf{r})\,\mathbf{E}_0\,dV = Q\mathbf{E}_0. \tag{3.8}$$

An external field will in general also exert a torque on a charge distribution:

$$\text{Torque} = \int\mathbf{r}\times(\rho(\mathbf{r})\,\mathbf{E}_0\,dV) = \mathbf{p}\times\mathbf{E}_0. \tag{3.9}$$

If $Q = 0$, there is *only* a torque, which then does not depend on the origin of coordinates; this torque tends to turn the charge distribution so that the dipole moment aligns with the field.

The tendency of the dipole moment to align with the field can also be understood by considering the energy of the charge distribution *in the external field*. The external potential can be written (Problem 2.1)

$$\Phi(\mathbf{r}) = \Phi(0) - \mathbf{E}_0 \cdot \mathbf{r}.$$

Hence (§2.3):

$$\text{potential energy} = \int \Phi(\mathbf{r})\rho(\mathbf{r})\,\mathrm{d}V$$

$$= Q\Phi(0) - \mathbf{p} \cdot \mathbf{E}_0. \tag{3.10}$$

Thus if the dipole is free to rotate the energy is a minimum when the dipole moment and external field are parallel.

Equation (3.8) shows there is no net force on a charge distribution in a uniform field if the total charge is zero. However, there can be a net force in a non-uniform field. Suppose the field varies sufficiently slowly over the charge distribution that it may be approximated by its first order Taylor expansion about an origin within the distribution; then

$$\mathbf{E}(\mathbf{r}) = \mathbf{E}(0) + x\frac{\partial \mathbf{E}}{\partial x} + y\frac{\partial \mathbf{E}}{\partial y} + z\frac{\partial \mathbf{E}}{\partial z}$$

$$= \mathbf{E}(0) + (\mathbf{r} \cdot \nabla)\mathbf{E}.$$

(The operator acts on each component of **E**, and the derivatives are evaluated at $\mathbf{r} = 0$.)

The total force is given by

$$\text{Force} = \int \rho(\mathbf{r})[\mathbf{E}(0) + (\mathbf{r} \cdot \nabla)\mathbf{E}]\,\mathrm{d}V$$

$$= Q\mathbf{E}(0) + (\mathbf{p} \cdot \nabla)\mathbf{E}, \tag{3.11}$$

and in general the second term is non-vanishing even if $Q = 0$.

The expressions we have obtained in this chapter are evidently valid both for charge distributions $\rho_{at}(\mathbf{r})$ defined on the atomic scale, and $\rho(\mathbf{r})$ defined on a macroscopic scale.

Problems

3.1 Show that the electric dipole moment of a charge distribution which is overall neutral, is independent of the origin of coordinates.

3.2 Derive the expression (3.5) for the electric field of a dipole. (You will need the result $\nabla(\mathbf{p} \cdot \mathbf{r}) = \mathbf{p}$.)

3.3 Show that the potential energy of two small dipoles $\mathbf{p}_1$ and $\mathbf{p}_2$, a distance **r** apart, is

$$U = \frac{1}{4\pi\varepsilon_0}\frac{1}{r^3}\left[\mathbf{p}_1 \cdot \mathbf{p}_2 - \frac{3(\mathbf{p}_1 \cdot \mathbf{r})(\mathbf{p}_2 \cdot \mathbf{r})}{r^2}\right].$$

3.4   A water molecule, approximated as a small dipole, is 2.5 Å distant from
       a singly charged cation (positive ion) in solution. What is the configuration
       of minimum potential energy of the water molecule in the field of the ion?
       In this configuration, is the force on the water molecule due to the cation
       field attractive or repulsive? What is the energy required to reverse the
       orientation of the water molecule? Compare this energy with $k_B T$ at room
       temperature. (The dipole moment of a water molecule is 6.17×10^{-30} C m.)

3.5   A charge distibution consists of two point charges q at $(a, 0, 0)$ and $(-a,
       0, 0)$, and two point charges $-q$ at $(0, a, 0)$ and $(0, -a, 0)$. Show that the
       distribution has no net charge and no dipole moment, but has a
       quadrupole moment such that $Q_{11} = 6a^2q$, $Q_{22} = -6a^2q$, and all other Q_{ij}
       are zero.

3.6   Show that the dipole moment of an atomic charge distribution is the same
       as that of the averaged charge distribution.

4

Static magnetic fields

Phenomena associated with magnets have been known for many centuries. The fact that a compass needle experiences a torque which aligns it in a particular direction was known to Chinese mariners before 1100, and in Europe about a century later; it was vital to the navigators of the great age of exploration. We now interpret this phenomenon as due to the interaction of the needle with the earth's magnetic field. The torque on the compass needle is similar to that on an electric dipole in an electric field; the needle behaves like a magnetic dipole and the fact that it experiences no net force tells us that it carries no net 'magnetic charge'.

Until 1819 the connection between magnetism and electricity was unknown, but in that year the Danish physicist Oersted observed that an electric current flowing in a wire deflected a nearby compass needle. Conversely, by Newton's law of action and reaction, the compass needle could be expected to exert forces on the current carrying wire. Oersted's discovery created great excitement in scientific academies throughout Europe and in particular stimulated the more detailed investigations of Biot and Savart, and of Ampère, in Paris. It was Ampère who found that two current carrying wires interacted by magnetic forces.

To account for the experimental phenomena it is natural to introduce a *magnetic field* $\mathbf{B}(\mathbf{r})$, which is determined by the magnets and current flows in the system under consideration, and through which different parts of the system interact.

4.1 The Lorentz force

Bearing in mind the modern technology of particle accelerators, electron microscopes and television sets, we choose as our starting point the influence of a magnetic field on a moving charged particle. Experimentally it is found that a particle carrying charge Q, and moving in a vacuum with velocity $\mathbf{v}$, experiences a force $\mathbf{F}$ which is given by

31

$$\blacktriangleright \qquad\qquad \mathbf{F} = Q(\mathbf{E} + \mathbf{v} \times \mathbf{B}), \qquad\qquad (4.1)$$

so that (apart from 'radiative corrections' which are usually small: see §17.4) an electron (say) of momentum $\mathbf{p}$ has the classical equation of motion

$$\frac{d\mathbf{p}}{dt} = (-e)(\mathbf{E} + \mathbf{v} \times \mathbf{B}). \qquad\qquad (4.2)$$

We regard (4.1) as an 'operational definition' of the field $\mathbf{B(r)}$ (cf. §2.2); in principle $\mathbf{B(r)}$ at a point $\mathbf{r}$ can be determined from observations of the force on moving charged test particles. In writing (4.1) without introducing another constant, the SI unit in which $\mathbf{B}$ is measured, the *tesla* (T), is determined, given the unit of charge.

The expression (4.1) is called the *Lorentz force*. The vector product $\mathbf{v} \times \mathbf{B}$ incorporates the fact that the force due to the $\mathbf{B}$ field is perpendicular both to the $\mathbf{B}$ field and to the velocity. Since $\mathbf{v} \cdot (\mathbf{v} \times \mathbf{B}) \equiv 0$, the magnetic field does no mechanical work on a moving charged particle (cf. Problem 4.4).

Although we take the Lorentz force on a particle to be the defining measure of a magnetic field there are many situations in which such a measurement would be inconvenient or impossible. We shall come later to other, more practical, methods of measuring $\mathbf{B(r)}$.

4.2 Gauss's theorem for the magnetic field

It was an early observation that magnetic fields, whether produced by electric currents or by magnets, are characterised by a dipole (or higher multipole) form at large distances. Despite intensive search, nobody has ever discovered an unambiguous source of magnetic flux (a 'magnetic monopole'). All the experimental evidence is consistent with the total *magnetic flux* through any *closed* surface being exactly zero:

$$\blacktriangleright \qquad\qquad \int_S \mathbf{B} \cdot d\mathbf{S} = 0. \qquad\qquad (4.3)$$

This is the magnetic equivalent of Gauss's theorem for the electric field (equation (2.18)). Applying the divergence theorem to the integral gives

$$\int_V \boldsymbol{\nabla} \cdot \mathbf{B} \, dV = 0.$$

Since this holds for any volume V, the integrand must vanish everywhere, so that

$$\blacktriangleright \qquad\qquad \boldsymbol{\nabla} \cdot \mathbf{B(r)} = 0. \qquad\qquad (4.4)$$

This is the magnetic equivalent of equation (2.17).

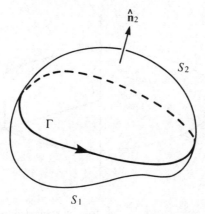

Fig. 4.1 Two surfaces S_1 and S_2 with the same boundary curve Γ. Together, S_1 and S_2 form a closed surface, through which the net magnetic flux is zero. Thus the flux inwards through S_1 equals the flux outwards through S_2. These flux directions (inwards through S_1, outwards through S_2) are determined by the sense of Γ indicated.

A consequence of (4.3) is that, for a given magnetic field, the flux through an *open* two-sided surface spanning a closed curve Γ is the same for any such surface (see Fig. 4.1).

4.3 Ampère's circuital law

The magnetic fields produced by steady currents flowing in thin wires were an early subject of investigation, and the laws governing them are well established. In particular, the principle of superposition holds for magnetic fields as it holds for electric fields: magnetic fields produced by different sources simply add vectorially. Very soon after Oersted's observations, Biot and Savart found the $(1/\rho)$ dependence of the magnetic field due to a steady current I flowing in a long straight wire of circular cross-section (Fig. 4.2). In cylindrical polar coordinates (ρ, ϕ, z), with the wire as z-axis, at points $\mathbf{r}$ outside the wire it is found that

$$\mathbf{B}(\mathbf{r}) = \frac{\mu_0}{2\pi} \frac{I}{\rho} \hat{\boldsymbol{\phi}}, \tag{4.5}$$

where μ_0 is a constant called the *permeability of free space*. We envisage the experiment as carried out in a vacuum (though the effect of air is very small). This expression holds at points sufficiently close to the wire for the field produced by other parts of the circuit to be negligible. The formula is probably well known to you!

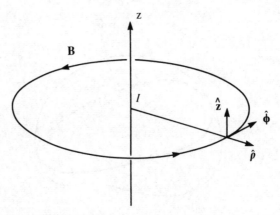

Fig. 4.2 The magnetic field due to a current I flowing in a long straight wire: lines of the **B** field encircle the wire.

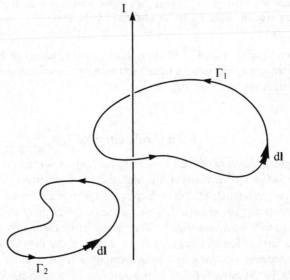

Fig. 4.3 Closed paths of integration Γ_1 and Γ_2 near a long straight current carrying wire. In cylindrical coordinates $d\mathbf{l} = d\rho\hat{\boldsymbol{\rho}} + \rho\,d\phi\hat{\boldsymbol{\phi}} + dz\hat{\mathbf{z}}$ and $\mathbf{B} = (\mu_0 I/2\pi\rho)\,\hat{\boldsymbol{\phi}}$.

Suppose we take a path Γ_1 which encloses the wire (Fig. 4.3) and calculate the line integral $\oint \mathbf{B} \cdot d\mathbf{l}$. Now $d\mathbf{l} = (d\rho, \rho\,d\phi, dz)$, using cylindrical polar coordinates, so that

$$\mathbf{B} \cdot d\mathbf{l} = \frac{\mu_0}{2\pi} I\,d\phi.$$

For the path $\Gamma_1, \oint_{\Gamma_1} d\phi = 2\pi$. Hence

$$\oint_{\Gamma_1} \mathbf{B} \cdot d\mathbf{l} = \mu_0 I. \tag{4.6}$$

For a path Γ_2 which does not enclose the wire, $\oint_{\Gamma_2} d\phi = 0$, giving

$$\oint_{\Gamma_2} \mathbf{B} \cdot d\mathbf{l} = 0. \tag{4.7}$$

It is remarkable that equations (4.6) and (4.7) are found to be true not only in the special case we have just considered, but for *any* closed path Γ threading through *any* circuits in which steady currents are flowing. In the general case, I is the total current which flows through a surface $S(\Gamma)$ bounded by Γ. (With steady current flows, it follows from the conservation of charge that I does not depend on the particular surface chosen.) The 'right-hand rule' linking the sense of the line integral and the direction of I is that explained in the Mathematical Prologue.

We can write

$$I = \int_S \mathbf{J} \cdot d\mathbf{S}, \tag{4.8}$$

where $\mathbf{J}$ is the current density in the wires (cf. §1.5) so that

$$\blacktriangleright \qquad \oint_\Gamma \mathbf{B} \cdot d\mathbf{l} = \mu_0 \int_S \mathbf{J} \cdot d\mathbf{S}. \tag{4.9}$$

We shall assume that equation (4.9) also holds for paths passing *through* a current distribution.

Stokes's theorem (P.8) applied to the left-hand side of (4.9) enables us to transform the line integral to an integral over the surface S, giving

$$\int_S \nabla \times \mathbf{B} \cdot d\mathbf{S} = \mu_0 \int_S \mathbf{J} \cdot d\mathbf{S}. \tag{4.10}$$

Since this is true for *all* surfaces S, it follows that the magnetic field $\mathbf{B}(\mathbf{r})$ produced by a distribution of steady current $\mathbf{J}(\mathbf{r})$ satisfies the field equation

$$\blacktriangleright \qquad \nabla \times \mathbf{B}(\mathbf{r}) = \mu_0 \mathbf{J}(\mathbf{r}). \tag{4.11}$$

Thus, given a field $\mathbf{B}(\mathbf{r})$, the current density $\mathbf{J}(\mathbf{r})$ which gives rise to it is determined. Conversely, the field equations (4.4) and (4.11) determine the magnetic field produced by a given distribution of steady currents $\mathbf{J}(\mathbf{r})$; this will be shown in Chapter 9, where the field equations are solved. Equation (4.11) is the magnetic equivalent of equation (2.10). Since the right-hand side of (4.11) does not vanish, we cannot *in general* express $\mathbf{B}(\mathbf{r})$ as the gradient of a potential function.

Our discussion and our notation has implied that the equations for the magnetic field are established for steady macroscopic current distributions. However, the validity of the equations at the atomic level is confirmed by experiment, and we assert that for steady atomic currents:

$$\nabla \cdot \mathbf{B}_{at} = 0, \quad \nabla \times \mathbf{B}_{at} = \mu_0 \mathbf{J}_{at}. \tag{4.12}$$

The field equations (4.4) and (4.11) follow from (4.12) on spatial averaging.

4.4 The magnetic force on a current carrying wire

The forces on a wire carrying a current I, at rest in an external magnetic field $\mathbf{B}$, follow from the Lorentz force law (4.1). Let us consider a wire sufficiently thin for a small (macroscopic) element of the wire to be specified by its length and direction $\delta\mathbf{l}$. We take the current density $\mathbf{J}$ in the element to be uniform, and then it must be in the direction of $\delta\mathbf{l}$. The current I is the integral of $\mathbf{J}$ over a cross-section $\mathbf{S}$ of the element (equation (1.16)). Hence

$$I\delta\mathbf{l} = (\mathbf{S} \cdot \mathbf{J})\,\delta\mathbf{l} = \mathbf{J}(\mathbf{S} \cdot \delta\mathbf{l}) = \mathbf{J}\delta V, \tag{4.13}$$

where $\delta V = \mathbf{S} \cdot \delta\mathbf{l}$ is the volume of the element.

Suppose that the charge carriers are electrons, with $Q = -e$. Then from §1.4, $\mathbf{J} = \rho_{el}\bar{\mathbf{v}}$, where ρ_{el} is the charge density of the conduction electrons and $\bar{\mathbf{v}}$ their mean velocity. Hence

$$\mathbf{J}\delta V = \rho_{el}\,\delta V\bar{\mathbf{v}} = \delta N(-e)\bar{\mathbf{v}},$$

where δN is the number of electrons in δV.

The total Lorentz force on these electrons is

$$\delta\mathbf{F} = \sum_i (-e)\mathbf{v}_i \times \mathbf{B};$$

$\mathbf{v}_i$ is the velocity of the ith electron. Since $\sum_i \mathbf{v}_i = \delta N\bar{\mathbf{v}}$, we obtain

$$\delta\mathbf{F} = (-e)\delta N\bar{\mathbf{v}} \times \mathbf{B} = (\mathbf{J}\delta V) \times \mathbf{B} = I\delta\mathbf{l} \times \mathbf{B}. \tag{4.14}$$

This force acting on the electrons is transmitted to the background lattice of ions, so that (4.14) is the force on the element of the wire. Note that $\delta\mathbf{F}$ is perpendicular to $d\mathbf{l}$. The total force on a section of the wire is found by integration:

$$\blacktriangleright \qquad \mathbf{F} = I\int d\mathbf{l} \times \mathbf{B}. \tag{4.15}$$

4.5 SI units (Système International)

It is only at this stage of our discourse that we can define the SI units of electricity and magnetism.

Consider the simple geometry of two thin parallel wires, distance p apart, carrying currents I_1 and I_2. Equations (4.5) and (4.15) show that, if the currents are in the same direction, the magnetic force per unit length on each wire is attractive and of magnitude

$$\frac{\mu_0}{2\pi} \frac{I_1 I_2}{p}.$$

Hence μ_0 has the dimensions kg m C^{-2}. In SI units, μ_0, the permeability of free space is defined to be

$$\mu_0 = 4\pi \times 10^{-7} \text{ kg m } C^{-2}.$$

The SI unit of current, the ampere, and the SI unit of charge, the coulomb ($1 \text{ A} = 1 \text{ C s}^{-1}$), are thereby determined. Thus two currents, each of 1 A, flowing in wires 1 m apart give a force per unit length of 2×10^{-7} N m^{-1}. This example indicates how in principle absolute standards of current (and hence charge) may be established in terms of the forces between circuits.

The relation between this definition of the ampere and coulomb, and the ε_0 introduced into Coulomb's law in §2.1, will become clear in Chapter 6.

Problems

4.1   An electron has velocity perpendicular to a uniform magnetic field $(0, 0, B)$. Show that it executes circles with an angular frequency $\omega = eB/m$. Calculate this *cyclotron frequency* for $B = 1$ T.

Show that, in general, the trajectory of an electron in a uniform magnetic field is a helix.

4.2   A particle of charge e and mass m moves in a region of uniform electric field E directed along Ox and a uniform magnetic field B directed along Oz. Show that its motion is given by the equations

$$\ddot{x} = (e/m)(E + \dot{y}B), \quad \ddot{y} = -(e/m)\dot{x}B, \quad \ddot{z} = 0.$$

Show that one solution (called a particular integral) is $x = 0$, $y = -Et/B$, $z = 0$, and that the general solution of the motion in the xy-plane is

$$x = R \sin \omega(t - t_0) + x_0,$$
$$y = R \cos \omega(t - t_0) + y_0 - Et/B,$$

where $\omega = eB/m$ is the cyclotron frequency.

Thus the particle executes circles superposed upon a drift velocity E/B perpendicular to both $\mathbf{E}$ and $\mathbf{B}$. The radius R and other constants are determined by the initial conditions.

4.3   A dilute solution of positive and negative ions flows with velocity v through a plastic pipe which is perpendicular to a constant magnetic field B. Explain why an electric potential difference is generated across the pipe, and estimate its magnitude if $v = 1$ m s^{-1} and $B = 1$ T, and the radius of the pipe is 1 cm.

4.4   A particle of mass m carrying charge Q moves in a static electromagnetic
        field according to Newton's law, $m(d\mathbf{v}/dt) = Q(\mathbf{E} + \mathbf{v} \times \mathbf{B})$.
            Obtain the energy equation

$$\tfrac{1}{2}mv^2 + Q\Phi(r) = \text{constant}$$

        where $\Phi(r)$ is the electrostatic potential.

4.5   The direction of the electron beam of Problem 1.6 can be changed by
        passing the beam through a region of transverse magnetic field. Estimate
        the magnitude of the magnetic field needed to deflect the beam through
        $10°$, if the field is uniform and extends over a distance of 4 cm.

4.6   Show that a radial magnetic field (i.e., a field of the form $B(r)\hat{\mathbf{r}}$) cannot
        exist.

4.7   A coaxial cable consists of a solid cylindrical conductor of radius a,
        surrounded by a conducting cylinder of inner radius b_1, outer radius b_2.
        The space between is filled with insulating material. A current I flows
        through the inner conductor, and returns through the outer conductor.
            Use Ampère's law to find expressions for the magnetic field at points in
        and between the conductors. Assume that the current density is uniform
        in each conductor, and all the materials are non-magnetic.

4.8   A uniform magnetic field $(B, 0, 0)$ is maintained in the region between the
        planes $z = 0$ and $z = a$, and the magnetic field is zero elsewhere.
            What current distribution produces the field?

4.9   The ends of a flexible conducting wire of radius a, length l and modulus
        of elasticity λ, are held at a distance l apart under no tension. The wire is
        in a magnetic field B perpendicular to its length.
            Show if a small current I passes through the wire then it becomes under
        tension

$$T = \frac{1}{2}\left(\frac{\pi\lambda}{3}\right)^{\frac{1}{3}}(IBla)^{\frac{2}{3}},$$

        and the centre is displaced a transverse distance

$$d = \frac{l}{4}\left(\frac{3IBl}{\pi a^2 \lambda}\right)^{\frac{1}{3}}.$$

        Estimate T and d if $I = 1$ A, $B = 1$ T, $l = 1$ m, $a = 1$ mm and $\lambda = 10^{11}$ N m^{-2}.

5

Time-dependent fields: Faraday's law and Maxwell's equations

Oersted's discovery of the magnetic effect of currents not only stimulated renewed interest in electricity and magnetism, but also led to the development of sensitive instruments which used the deflection of magnets to measure currents; a simple galvanometer was devised by Schweigger in 1820, and the more sensitive astatic galvanometer by Nobili in 1825. Previously, electric currents could only be detected by the observation of sparks, or by their chemical effects in electrolysis. The latter method was quantitative, but not well suited to the detection of small currents. The use of the galvanometer was important in the experimental work of Faraday at the Royal Institution in London, where in 1831–2 he carried out a now famous series of experiments on the induction of electric currents by magnetic fields.

Faraday found that a current was induced to flow round a closed conducting circuit when a nearby magnet was moved, or the current in a nearby circuit was changed, or when the circuit was moved in a fixed magnetic field. In all these cases he established that the induced current was proportional to the rate of change of magnetic flux through the circuit.

5.1 Faraday's law of induction

The current flow that Faraday observed when a coil of wire was moving in a static magnetic field $\mathbf{B}(\mathbf{r})$ can be understood in terms of effects we have already discussed.

The charge carriers in the element $\delta\mathbf{l}$ of the circuit are subject to an additional Lorentz force $\mathbf{F} = Q\mathbf{u} \times \mathbf{B}$ due to the velocity $\mathbf{u}$ of that element. The component of $\mathbf{F}$ along $\delta\mathbf{l}$ drives mobile conduction electrons ($Q = -e$)

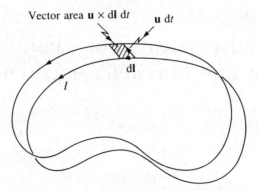

Fig. 5.1 An element dl of a moving circuit sweeps out an area dS = $\mathbf{u}\,dt \times dl$ in time dt. The magnetic flux through the circuit at time $t+dt$ differs from that at time t by

$$d\mathscr{F} = \mathscr{F}(t+dt) - \mathscr{F}(t) = \sum_{\text{strip}} \mathbf{B} \cdot (\mathbf{u} \times dl)\,dt$$

$$= \oint_{\Gamma} \mathbf{B} \cdot \mathbf{u} \times dl\,dt.$$

The direction of dS with respect to the direction of dl is determined by the right-hand rule explained in the Mathematical Prologue. The circuit need not be rigid. The velocity $\mathbf{u}$ may vary round the circuit.

round the circuit to produce the current. The induced *electromotive force*, or e.m.f., is defined to be the line integral of $(\mathbf{F}/Q)$ round the circuit:

▶
$$\text{e.m.f.} = \mathscr{E} = \oint (\mathbf{F}/Q) \cdot \delta l.$$

(The term e.m.f. is a misnomer, since $\mathscr{E}$ has the dimensions of potential, but it is long established.)

Since $\mathbf{F} = Q\mathbf{u} \times \mathbf{B}$, we have

$$\mathscr{E} = \oint \mathbf{u} \times \mathbf{B} \cdot \delta l$$

$$= -\oint \mathbf{B} \cdot \mathbf{u} \times \delta l \quad \text{(triple scalar product!)}$$

$$= -\frac{d\mathscr{F}}{dt}, \tag{5.1}$$

where $\mathscr{F}(t)$ is the magnetic flux through the circuit at time t (see Fig. 5.1). The minus sign in this equation implies 'Lenz's law': the direction of induced current flow in the circuit is such that the magnetic flux it produces opposes the change in flux through the circuit. The SI unit of magnetic flux is the *weber* (Wb).

That the same current should flow in this coil when it is held at rest, and the magnet producing the field is moved to give the same relative motion of coil and magnet, will come as no surprise to the reader brought up on the principle of equivalence of inertial frames of reference. However, the interpretation of the effect is very different: in a circuit at rest there is no 'additional Lorentz force', and we must conclude that when the magnetic flux through a circuit is changing, an actual electric field $E(r)$ is induced, giving a force $F = QE(r)$ on a charge carrier. Since the induced current is the same, the e.m.f. must be the same, so that

$$\mathscr{E} = \oint (F/Q) \cdot dl = \oint E \cdot dl = -\frac{d\mathscr{F}}{dt}, \tag{5.2}$$

where now

$$\frac{d\mathscr{F}}{dt} = \frac{d}{dt} \int_S B \cdot dS = \int_S \frac{\partial B}{\partial t} \cdot dS. \tag{5.3}$$

The last step is valid since the circuit is held fixed.

Equations (5.1) and (5.2) express, in our notation, the relations found experimentally by Faraday. We emphasise that these are two independent phenomena. The first, which is based on the Lorentz force law, requires the presence of a moving circuit. In the second case, when the magnetic field is changing and the circuit is fixed, there is an induced electric field, satisfying

$$\blacktriangleright \qquad \oint E \cdot dl = -\int \frac{\partial B}{\partial t} \cdot dS. \tag{5.4}$$

This line integral of the electric field is manifestly independent of the material of the wire and can be taken to exist even if a material circuit is not present.

(However, the induced fields E_1 and E_2 in the presence and absence of a conducting circuit will in general differ, since an induced current flow in the circuit, as well as producing a magnetic field which contributes to the flux (§9.4), will also redistribute charge density on the wire (§8.5).)

We can convert (5.4) into a field equation, just as we converted Ampère's circuital law into a field equation. Using Stokes's theorem, we obtain

$$\int_S \nabla \times E \cdot dS = -\int_S \frac{\partial B}{\partial t} \cdot dS.$$

Since this equation is taken to hold for *all* surfaces S it follows that

$$\blacktriangleright \qquad \nabla \times E + \frac{\partial B}{\partial t} = 0. \tag{5.5}$$

This is Faraday's law of induction in differential form. It replaces equation (2.10) of electrostatics, and one consequence of Faraday's law is that $\mathbf{E}$ cannot in general be expressed as the gradient of a scalar potential.

5.2 Maxwell's equations

Let us set out the equations for the macroscopic electric and magnetic fields that we have so far assembled. In local form they are:

$$
\begin{array}{llll}
\text{Coulomb's law} & \nabla \cdot \mathbf{E} = \rho/\varepsilon_0 & \text{(a)} & \\
\text{Ampère's law} & \nabla \times \mathbf{B} = \mu_0 \mathbf{J} & \text{(b)} & \\
\text{The absence of magnetic poles} & \nabla \cdot \mathbf{B} = 0 & \text{(c)} & \text{(5.6)} \\
\text{Faraday's law of induction} & \nabla \times \mathbf{E} + \dfrac{\partial \mathbf{B}}{\partial t} = 0 & \text{(d)} &
\end{array}
$$

The charge density ρ and the current density $\mathbf{J}$ are related by the continuity equation (1.15):

$$
\frac{\partial \rho}{\partial t} + \nabla \cdot \mathbf{J} = 0. \quad \text{(e)}
$$

Equations (a), (b), and (c) have been derived from experiments performed in steady state situations where fields and currents are not changing with time. Maxwell, who was largely responsible for putting these laws in the form of differential equations, contemplated the possibility that they were all valid even in time-dependent situations, and realised that Ampère's law (b) was inconsistent with the continuity equation. Because $\nabla \cdot \nabla \times \mathbf{B} = 0$ is a mathematical identity for *any* field $\mathbf{B}$, if we take the divergence of (b) we obtain $\nabla \cdot \mathbf{J} = 0$.

Although this is true for a steady state – in which Ampère's law obtains – it is false when the charge density is changing with time. Maxwell saw that if Ampère's law is modified by the addition of a time derivative (which leaves it unaltered in a steady state), to become

$$
\nabla \times \mathbf{B} - \mu_0 \varepsilon_0 \frac{\partial \mathbf{E}}{\partial t} = \mu_0 \mathbf{J},
$$

then we shall have a set of equations which are mathematically consistent even for $\rho(\mathbf{r}, t)$ and $\mathbf{J}(\mathbf{r}, t)$ varying with time. In particular, taking the divergence of the modified equation, and using (a), leads now to the continuity equation, since $\nabla \cdot (\partial \mathbf{E}/\partial t) = \partial(\nabla \cdot \mathbf{E})/\partial t = \partial(\rho/\varepsilon_0)\partial t$. (The expression $\varepsilon_0 \, \partial \mathbf{E}/\partial t$ has the dimensions of current density and was called by Maxwell the 'displacement current', a term still occasionally used.)

We replace the set of equations (5.6) by *Maxwell's equations*:

$$\nabla \cdot \mathbf{E} = \rho/\varepsilon_0 \quad \text{(a)}$$

$$\nabla \times \mathbf{B} - \mu_0 \varepsilon_0 \frac{\partial \mathbf{E}}{\partial t} = \mu_0 \mathbf{J} \quad \text{(b)}$$

▶ $$\hspace{6cm} (5.7)$$

$$\nabla \cdot \mathbf{B} = 0 \quad\quad \text{(c)}$$

$$\nabla \times \mathbf{E} + \frac{\partial \mathbf{B}}{\partial t} = 0 \quad\quad \text{(d)}$$

The continuity equation is now redundant, since it is a consequence of (a) and (b).

If $\rho(\mathbf{r}, t)$ and $\mathbf{J}(\mathbf{r}, t)$ are known as functions of $\mathbf{r}$ and t, then Maxwell's equations may be solved for the fields $\mathbf{E}(\mathbf{r}, t)$ and $\mathbf{B}(\mathbf{r}, t)$, as we shall see in Chapter 16. More generally, of course, ρ and $\mathbf{J}$ are themselves functions of $\mathbf{E}$ and $\mathbf{B}$, since electric charges are driven by the Lorentz force

$$\mathbf{F} = Q(\mathbf{E} + \mathbf{v} \times \mathbf{B}).$$

In solving practical problems involving material media, additional phenomenological equations are usually introduced to describe the response of the charges in the materials to the fields. We shall discuss these in later chapters.

Maxwell developed his ideas in a series of papers between 1861 and 1868. Subsequent experimental and theoretical investigations have demonstrated a remarkable range of applicability of the theory. We have written them out for macroscopic fields and macroscopic ρ and $\mathbf{J}$, but they hold on atomic and nuclear length and time scales, and (5.7) can be regarded as obtained by averaging the more basic equations.

Maxwell's equations encompass light waves and the phenomena of optics; they turn out to be consistent with Einstein's special theory of relativity; in 1927 they were put in quantum form by Dirac. In the following chapters we shall try to exhibit something of all this.

Problems

5.1   Electric power generators convert mechanical power into electrical power; they are an indispensable part of our technology. To illustrate their working principle, consider a plane loop of wire of area S rotating in a uniform magnetic field $\mathbf{B}$ about an axis in its plane perpendicular to the field, with angular velocity ω. If at time $t = 0$ the loop is also perpendicular to $\mathbf{B}$, show that the flux through the loop at any time is $\mathscr{F}(t) = BS \cos \omega t$, and the e.m.f. $= BS\omega \sin \omega t$.

5.2   If the circuit of Problem 5.1 is closed and has resistance R, then a current $I = (\text{e.m.f.})/R$ flows, and the electrical power produced is directly converted into Joule heating $= I^2 R = (BS\omega)^2 \sin^2 \omega t/R$. Show that this is equal to the rate of work of the external forces acting on the wire that are necessary to maintain rotation. (Take a rectangular loop for simplicity.)

5.3   A time-dependent magnetic field $(0, 0, B(\rho, t))$ has cylindrical symmetry about the z-axis. Show that the resulting electric field lines are circles about the z-axis, and the field has magnitude

$$E(\rho, t) = -\frac{1}{2\pi\rho}\frac{d\mathscr{F}}{dt}$$

where $\mathscr{F}(\rho, t)$ is the magnetic flux at time t through the circles of radius ρ.

If an electron is constrained to move on a circle of radius ρ show that, starting from rest when the **B** field is zero, it will after time t acquire momentum $(e/2\pi\rho)\mathscr{F}(\rho, t)$.

Show that the magnetic field on the circle will itself constrain the electron to the circle, if it is one half of the average field through the circle.

This example illustrates the principle of the *betatron*, a machine for accelerating electrons to high energies. If $\rho = 1$ m, what is the electron momentum when $\mathscr{F} = 1$ T m^{-2}? Express your answer in MeV/c (1 eV = 1.6×10^{-19} J), and compare with $m_e c^2/c$ where $m_e = 0.51$ MeV/c^2 is the electron mass.

5.4   A radioactive sphere with a radius of 1 cm ejects 10^{10} electrons per second, so that the averaged macroscopic current outside the sphere is radial and has spherical symmetry. Is there a macroscopic magnetic field? How is the Maxwell equation (5.7b) satisfied?

5.5   Show that if the combination of fields $\mathbf{E}(\mathbf{r}, t)$ and $\mathbf{B}(\mathbf{r}, t)$ is a solution of Maxwell's equations in free space, so is the combination $\mathbf{E}'(\mathbf{r}, t)$, $\mathbf{B}'(\mathbf{r}, t)$, where $\mathbf{E}'(\mathbf{r}, t) = -c\mathbf{B}(\mathbf{r}, t)$, $\mathbf{B}'(\mathbf{r}, t) = (1/c)\mathbf{E}(\mathbf{r}, t)$.

6

Electromagnetic waves in a vacuum

6.1 Maxwell's equations in a vacuum

In Chapter 5 we formulated Maxwell's equations. We shall now look for solutions of these equations in a particularly simple case: we consider a region of space which is empty of everything except electric and magnetic fields. Maxwell's equations (5.7) reduced to

$$\nabla \cdot \mathbf{E} = 0 \quad \text{(a)}, \quad \nabla \times \mathbf{B} - \mu_0 \varepsilon_0 \frac{\partial \mathbf{E}}{\partial t} = 0 \quad \text{(b)},$$

$$\nabla \cdot \mathbf{B} = 0 \quad \text{(c)}, \quad \nabla \times \mathbf{E} + \frac{\partial \mathbf{B}}{\partial t} = 0 \quad \text{(d)}.$$

(6.1)

In a vacuum these equations, so far as is known, hold on all length scales, apart from quantum corrections which become significant only in very intense fields.

Taking the curl of (d), the vector identity

$$\nabla \times (\nabla \times \mathbf{E}) = \nabla (\nabla \cdot \mathbf{E}) - \nabla^2 \mathbf{E}$$

gives, with (a),

$$-\nabla^2 \mathbf{E} + \nabla \times \frac{\partial \mathbf{B}}{\partial t} = 0.$$

The ∇ and $\partial/\partial t$ operations can be interchanged, so that using (b) we obtain

$$\blacktriangleright \qquad \nabla^2 \mathbf{E} = \mu_0 \varepsilon_0 \frac{\partial^2 \mathbf{E}}{\partial t^2}.$$

(6.2)

Similarly, taking the curl of (b) and using (c) and (d), gives

$$\blacktriangleright \qquad \nabla^2 \mathbf{B} = \mu_0 \varepsilon_0 \frac{\partial^2 \mathbf{B}}{\partial t^2}.$$

(6.3)

Thus both the **E** and **B** fields satisfy the *wave equation*, with the wave velocity $(\mu_0 \varepsilon_0)^{-\frac{1}{2}}$, and we shall display below some particular wave solutions of Maxwell's equations. Transposing from the nineteenth century into SI units, Maxwell had available moderately accurate measurements of ε_0 and of the velocity of light c. Noting that the numerical value of $(\mu_0 \varepsilon_0)^{-\frac{1}{2}}$ (which indeed has the dimensions of velocity) was consistent with the values found for the velocity of light, he concluded that light was an electromagnetic phenomenon, and $c = (\mu_0 \varepsilon_0)^{-\frac{1}{2}}$. This successful identification, along with the prediction of an entire spectrum of electromagnetic radiation, is perhaps the greatest triumph of all nineteenth-century theoretical physics.

Indeed, from 1983, the velocity of light c together with the unit of time has been taken to define the unit of length: the metre is such that c in a vacuum is *exactly* $2.997\,924\,58 \times 10^8$ m s^{-1}. Since $\mu_0 = 4\pi \times 10^{-7}$ N A^{-2} and $c^2 = (\mu_0 \varepsilon_0)^{-1}$, it follows that the value of $(4\pi\varepsilon_0)^{-1}$ is also exactly defined: $(4\pi\varepsilon_0)^{-1} = 10^{-7}(c/\text{m s}^{-1})^2$ N C^{-2} m^2. Hence Coulomb's law (2.2) determines the unit of charge exactly; that this is identical with the unit of charge as defined from the force between current carrying wires (§4.5) is a consequence of the consistency of the overall theory.

6.2 Plane electromagnetic waves

A solution of the wave equation (6.2) which corresponds to a travelling wave in the Oz-direction, with plane wavefronts parallel to the Oxy-plane, is of the form

$$\mathbf{E}(z-ct) = [f(z-ct), g(z-ct), h(z-ct)],$$

where f, g, h can be any functions of $(z-ct)$, describing a pulse or a continuous wave as we choose, and $c = (\mu_0 \varepsilon_0)^{-\frac{1}{2}}$.

Although Maxwell's equations imply that the fields must satisfy the wave equation, they do not allow every solution of the wave equation; in particular, to satisfy $\mathbf{\nabla} \cdot \mathbf{E} = 0$ requires $\partial h(z-ct)/\partial z = 0$, giving $h = $ constant. A constant component of the field is of no interest here, so we take

$$\mathbf{E} = [f(z-ct), g(z-ct), 0]. \tag{6.4}$$

Then the Maxwell equation (6.1d) gives

$$\partial \mathbf{B}/\partial t = -\mathbf{\nabla} \times \mathbf{E} = -(0, 0, \partial/\partial z) \times \mathbf{E}(z-ct)$$
$$= [g'(z-ct), -f'(z-ct), 0].$$

Integrating with respect to t, the magnetic field associated with the wave is

$$c\mathbf{B} = [-g(z-ct), f(z-ct), 0]. \tag{6.5}$$

Clearly $\mathbf{\nabla} \cdot \mathbf{B} = 0$; the remaining Maxwell equation (6.1b) is also satisfied.

Thus the electric wave is accompanied by a magnetic wave, which we can write as

▶ $$c\mathbf{B} = \hat{\mathbf{z}} \times \mathbf{E}. \tag{6.6}$$

$\mathbf{E}$, $\mathbf{B}$ and the direction of propagation $\hat{\mathbf{z}}$ make up a right-handed set of orthogonal vectors. Everywhere $|\mathbf{E}| = c|\mathbf{B}|$; this means that $|\mathbf{B}|$, measured in teslas, is smaller numerically by eight orders of magnitude than $|\mathbf{E}|$, measured in volts per metre. It is important to realise that this is just an artefact of the SI system of units. In a plane wave there is in fact exactly as much energy in the magnetic field as in the electric field: we shall see later in the chapter that the energy density of an electromagnetic field $U(\mathbf{r}, t)$ is given by

$$U(\mathbf{r}, t) = \frac{\varepsilon_0}{2} |\mathbf{E}|^2 + \frac{1}{2\mu_0} |\mathbf{B}|^2$$

$$= \frac{\varepsilon_0}{2} (|\mathbf{E}|^2 + c^2 |\mathbf{B}|^2). \tag{6.7}$$

Since $|\mathbf{E}| = c|\mathbf{B}|$ in a plane wave, the two terms in this expression contribute equally.

6.3 Monochromatic plane waves and polarisation

A strictly monochromatic plane wave has everywhere a harmonic time dependence, and in this case (6.4) specialises to

$$\mathbf{E}(z - ct) = [E_1 \cos(kz - \omega t - \phi_1), E_2 \cos(kz - \omega t - \phi_2), 0], \tag{6.8}$$

where ω is the angular frequency, and $k = 2\pi/\lambda$ where λ is the wavelength; k and ω are related by the 'dispersion relation':

$$\omega = ck. \tag{6.9}$$

E_1, E_2 and ϕ_1, ϕ_2 are the (arbitrary) amplitudes and phases of the x and y components. The $\mathbf{B}$ field is then determined by (6.6): $c\mathbf{B}$ has everywhere, and at every instant, the same magnitude as $\mathbf{E}$ but is rotated by $\pi/2$ in a positive sense about the direction of propagation Oz.

The state of *polarisation* of a monochromatic plane wave is determined by the relative magnitudes of E_1 and E_2 and the phase difference $\phi = \phi_2 - \phi_1$. We may take any fixed point in space as origin $z = 0$, and choose the origin of t so that $\phi_1 = 0$. Then at that point

$$\mathbf{E}(t) = [E_1 \cos \omega t, E_2 \cos(\omega t + \phi), 0]. \tag{6.10}$$

If $\phi = 0$, $\mathbf{E}(t) = (E_1, E_2, 0) \cos \omega t$. In this case, $\mathbf{E}$ oscillates harmonically along a fixed line in space. The wave is said to be *linearly polarised* along that line: the $\mathbf{E}$ vectors at all points, at every instant, are parallel to $(E_1, E_2, 0)$ (Fig. 6.1). The less appropriate phrase 'plane polarised' is also used.

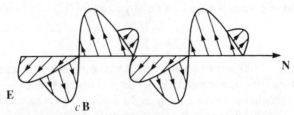

Fig. 6.1 A linearly polarised harmonic wave: the figure indicates the spatial dependence of the **E** and **B** fields at some instant along a line in space perpendicular to a wavefront. The profiles move with velocity c along this line, which is also the direction of the Poynting vector **N**. At any point, **E**, **B**, and **N** form a right-handed set.

If $\phi = -\pi/2$ and $E_1 = E_2 = E$,

$$\mathbf{E}(t) = (E\cos\omega t, E\sin\omega t, 0),$$

so that the electric field vector has a constant magnitude and rotates in a positive sense around the direction of propagation. The wave is said to be *circularly polarised*. If $\phi = \pi/2$ and $E_1 = E_2$ the wave is again circularly polarised, but the **E** vector rotates in the negative sense.

In the general case, the wave is said to be *elliptically polarised*. If, for example, $E_1 \neq E_2$, and $\phi = -\pi/2$,

$$\mathbf{E}(t) = (E_1\cos\omega t, E_2\sin\omega t, 0),$$

so that $\mathbf{E}(t) = (E_x, E_y, 0)$ lies on the ellipse

$$\frac{E_x^2}{E_1^2} + \frac{E_y^2}{E_2^2} = 1,$$

and the **E** vector rotates in a positive sense around the direction of propagation. In our example the axes of the ellipse are the axes of coordinates. In the general case **E** lies on an ellipse, the axes of which can have any orientation in the Oxy-plane.

Circularly polarised and elliptically polarised waves can be regarded as the superposition of two plane polarised waves – in our analysis the E_x and E_y components. Conversely, we can regard a plane polarised wave as the superposition of waves of positive and negative circular (or elliptic) polarisation; for example,

$$(E_1\cos\omega t, E_2\sin\omega t, 0) + (E_1\cos\omega t, -E_2\sin\omega t, 0) = (2E_1\cos\omega t, 0, 0).$$

The state of polarisation of an electromagnetic wave is of considerable physical significance, and technical importance.

6.4 Plane harmonic waves: the general case

In the last section we have presented a complete description of strictly monochromatic plane waves moving in the z-direction. More generally, a plane wave moving in a direction specified by a unit vector $\hat{\mathbf{k}}$ has as wavefronts the planes $\hat{\mathbf{k}} \cdot \mathbf{r} = $ constant, so that $\cos(kz - \omega t + \phi)$ above is replaced by $\cos(\mathbf{k} \cdot \mathbf{r} - \omega t + \phi)$, where

$$\mathbf{k} = k\hat{\mathbf{k}} = \frac{2\pi}{\lambda}\hat{\mathbf{k}} \tag{6.11}$$

is called the *wave vector*. The electric field $\mathbf{E}$ is perpendicular to $\mathbf{k}$, so that $\mathbf{k} \cdot \mathbf{E} = 0$, and (6.6) becomes

$$c\mathbf{B} = \hat{\mathbf{k}} \times \mathbf{E}. \tag{6.12}$$

It is often convenient to describe a plane wave as

$$\mathbf{E} = \text{Re}\,\mathbf{E}_0\,e^{i(\mathbf{k} \cdot \mathbf{r} - \omega t)}. \tag{6.13}$$

As long as only linear operations are involved, we can work with the complex field and omit 'Re', on the understanding that the real part is eventually to be used. For the plane wave (6.13), $\nabla \cdot \mathbf{E} = i\mathbf{k} \cdot \mathbf{E}$, $\nabla \times \mathbf{E} = i\mathbf{k} \times \mathbf{E}$, $\partial \mathbf{E}/\partial t = -i\omega\mathbf{E}$; it is easily seen directly that Maxwell's equations are satisfied provided that $\omega = ck$, $\mathbf{k} \cdot \mathbf{E}_0 = 0$, and $c\mathbf{B} = \hat{\mathbf{k}} \times \mathbf{E}$. Mathematical operations on the complex field are usually simpler to perform than the same operations on the real field.

Since Maxwell's equations in a vacuum are linear in the fields, any solutions may be superposed. It follows from Fourier's theorem that the most general wave patterns of travelling or standing waves can be constructed from plane waves of the form (6.13).

Electromagnetic radiation has been observed over a great range of frequencies, of which visible light is only a tiny fraction. Fig. 6.2 correlates frequencies, wavelengths, and photon energies in electron volts derived from the Planck relation $E = h\nu$, for the various bands of frequencies it has been found convenient to distinguish.

6.5 Electromagnetic field energy, and Poynting's vector

We showed in §2.6 that the electrostatic energy of a system of charged particles at rest could be regarded as residing entirely in the electric field, with an energy density in free space $\varepsilon_0\,\mathbf{E}^2/2$. We shall now consider general time-dependent electromagnetic fields, but restrict the discussion to free space.

For such a field, consider

$$\frac{\partial}{\partial t}\left(\frac{\varepsilon_0\,\mathbf{E}^2}{2}\right) = \varepsilon_0\,\mathbf{E} \cdot \frac{\partial \mathbf{E}}{\partial t}$$

$$= \frac{1}{\mu_0}\mathbf{E} \cdot \nabla \times \mathbf{B}, \text{ using the Maxwell equation (6.1b).}$$

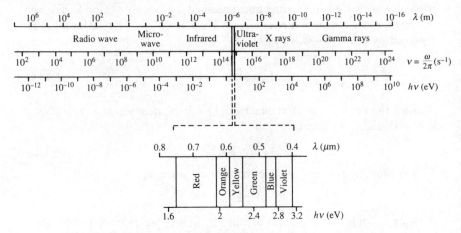

Fig. 6.2 The electromagnetic spectrum.

Then the vector identity $\nabla \cdot (\mathbf{E} \times \mathbf{B}) = \mathbf{B} \cdot \nabla \times \mathbf{E} - \mathbf{E} \cdot \nabla \times \mathbf{B}$, together with the Maxwell equation (6.1d) for $\nabla \times \mathbf{E}$, give

$$\frac{\partial}{\partial t}\left(\frac{\varepsilon_0 \mathbf{E}^2}{2}\right) = -\frac{1}{\mu_0}\nabla \cdot (\mathbf{E} \times \mathbf{B}) - \frac{1}{\mu_0}\mathbf{B} \cdot \frac{\partial \mathbf{B}}{\partial t}.$$

We rewrite this as

$$\frac{\partial}{\partial t}\left(\frac{\varepsilon_0 \mathbf{E}^2}{2} + \frac{\mathbf{B}^2}{2\mu_0}\right) + \nabla \cdot \left(\frac{\mathbf{E} \times \mathbf{B}}{\mu_0}\right) = 0. \tag{6.14}$$

This equation is of the same mathematical form as equation (1.15) which expresses charge conservation, and we can interpret it as expressing energy conservation in free space. Integrating (6.14) over any fixed volume V enclosed by a surface S, and using the divergence theorem on the last term, we obtain

$$\frac{\partial}{\partial t}\int_V \left(\frac{\varepsilon_0 \mathbf{E}^2}{2} + \frac{\mathbf{B}^2}{2\mu_0}\right)\mathrm{d}V + \int_S \left(\frac{\mathbf{E} \times \mathbf{B}}{\mu_0}\right) \cdot \mathrm{d}\mathbf{S} = 0. \tag{6.15}$$

The expression in parentheses in the first term can be interpreted as the energy density of the electromagnetic field, generalising the electrostatic case: here $\mathbf{B}^2/2\mu_0$ gives the energy density of the magnetic field. Thus the first term gives the rate of change of electromagnetic field energy inside the volume V. We interpret the integral over the surface S as the rate of flow of electromagnetic energy out of the volume. The vector

$$\blacktriangleright \qquad \mathbf{N} = \mu_0^{-1}\mathbf{E} \times \mathbf{B} \tag{6.16}$$

is called the *Poynting vector*. $\mathbf{N} \cdot \mathrm{d}\mathbf{S}$ gives the rate of flow of energy through the surface element dS. Overall, (6.15) expresses the conservation of electromagnetic energy in free space.

It is instructive to construct the Poynting vector for the general plane wave described by (6.4) and (6.5):

$$\mathbf{N} = \mu_0^{-1}(\mathbf{E} \times \mathbf{B}) = (c\mu_0)^{-1}(f^2 + g^2)\,\hat{\mathbf{z}}.$$

But the energy density in the wave is $\varepsilon_0(f^2 + g^2)$, and $(\varepsilon_0 \mu_0)^{-1} = c^2$, so that in this case

$$\mathbf{N} = cU(z - ct)\,\hat{\mathbf{z}},$$

and we see (as we would expect) that the wave is transporting its energy with velocity c.

6.6 Standing waves in cavities

As is usually the case with wave motion, we can construct standing wave solutions of Maxwell's equations by superposing appropriate travelling wave solutions. The ability to produce and sustain standing wave patterns in cavities is of considerable technological importance, not only in the culinary sciences (microwave ovens), but in all branches of science and engineering that use electromagnetic radiation at microwave frequencies. For example, in radar the microwave beam is derived from a cavity in which a standing wave pattern is maintained. We shall see that only certain frequencies are allowed for such standing waves, and the cavity can be tuned to a particular frequency by changing its geometry.

To illustrate this we consider the standing wave patterns possible in a cubical box, which we take to be enclosed by the planes $x = 0$, $x = L$; $y = 0$, $y = L$; $z = 0$, $z = L$. We must also specify the boundary conditions to be satisfied by the field: we shall require that the $\mathbf{E}$ field at a surface of the enclosure is normal to that surface. These conditions would apply if the box were constructed from a 'perfect conductor', and are a good approximation for, say, copper. We shall consider the boundary conditions at a metallic surface in detail in Chapter 13 and Appendix C.

A standing wave pattern for the electric field inside the cube which satisfies the boundary conditions is

$$E_x = \alpha_x \cos\left(\frac{l\pi x}{L}\right) \sin\left(\frac{m\pi y}{L}\right) \sin\left(\frac{n\pi z}{L}\right) \cos \omega t,$$

$$E_y = \alpha_y \sin\left(\frac{l\pi x}{L}\right) \cos\left(\frac{m\pi y}{L}\right) \sin\left(\frac{n\pi z}{L}\right) \cos \omega t, \qquad (6.17)$$

$$E_z = \alpha_z \sin\left(\frac{l\pi x}{L}\right) \sin\left(\frac{m\pi y}{L}\right) \cos\left(\frac{n\pi z}{L}\right) \cos \omega t,$$

where the amplitudes α_x, α_y, α_z are constants, and l, m, n are integers:

$l = 0, 1, 2, \ldots$; $m = 0, 1, 2, \ldots$; $n = 0, 1, 2, \ldots$. Negative integers merely correspond to changes of sign in the α_i, and do not give different patterns. If more than one of l, m, n is zero, the field vanishes identically.

Inside the cavity the field must satisfy Maxwell's equations. The wave equation (6.2) is satisfied by each component of $\mathbf{E}$ if

$$\frac{\omega^2}{c^2} = \frac{\pi^2}{L^2}(l^2 + m^2 + n^2), \qquad (6.18)$$

but this is not a sufficient condition. The Maxwell equation $\nabla \cdot \mathbf{E} = 0$ is only satisfied if

$$l\alpha_x + m\alpha_y + n\alpha_z = 0.$$

If one of the integers l, m, n vanishes, there is only one independent amplitude (see (6.17)), but otherwise two amplitudes are independent for each set (l, m, n). (Compare with the two independent states of polarisation of a monochromatic plane wave.)

The Maxwell equation $\partial \mathbf{B}/\partial t = -\nabla \times \mathbf{E}$ gives the magnetic field inside the cavity:

$$B_x = \beta_x \sin\left(\frac{l\pi x}{L}\right)\cos\left(\frac{m\pi y}{L}\right)\cos\left(\frac{n\pi z}{L}\right)\sin \omega t$$

$$B_y = \beta_y \cos\left(\frac{l\pi x}{L}\right)\sin\left(\frac{m\pi y}{L}\right)\cos\left(\frac{n\pi z}{L}\right)\sin \omega t$$

$$B_z = \beta_z \cos\left(\frac{l\pi x}{L}\right)\cos\left(\frac{m\pi y}{L}\right)\sin\left(\frac{n\pi z}{L}\right)\sin \omega t$$

where

$$\beta_x = (\alpha_y n - \alpha_z m)\,\pi/L\omega, \quad \beta_y = (\alpha_z l - \alpha_x n)\,\pi/L\omega,$$
$$\beta_z = (\alpha_x m - \alpha_y l)\,\pi/L\omega.$$

The two remaining Maxwell equations (6.1(b)) and (6.1(c)) are then (remarkably) satisfied identically, as too are the conditions for the $\mathbf{B}$ field on the surfaces of the cavity (Appendix C).

Problems

6.1   A one milliwatt laser emits a beam of (nearly) monochromatic light of circular cross-section with diameter 1 mm. Calculate the maximum electric and magnetic field strengths if the light is (a) linearly polarised, (b) circularly polarised.

6.2   Using the notation of §6.2, a plane electromagnetic pulse is described by

$$f(z - ct) = E_0\, e^{-(z-ct)^2/4a^2} \cos k_0(z - ct)$$
$$g(z - ct) = 0.$$

Sketch the wave form $f(z)$ at $t = 0$, if $a \gg 2\pi/k_0$.

Using the formula of Fourier transform theory,

$$\mathrm{e}^{-q^2/4a^2} = \frac{a}{\sqrt{\pi}} \int_{-\infty}^{\infty} \mathrm{e}^{-a^2 p^2} \cos(pq) \, \mathrm{d}p,$$

express this pulse as a superposition of monochromatic linearly polarised plane waves. What band of frequencies dominates this superposition?

6.3   An electron is subject to a plane polarised electromagnetic field with $\mathbf{E} = (E_0 \cos(kz - \omega t), 0, 0)$. The electron is at rest at the origin at $t = 0$. Show that, neglecting the force due to the magnetic field of the wave, it moves harmonically parallel to the x-axis with maximum velocity $eE_0/m_e \omega$.

Suppose that the wave is that of the laser of Problem 6.1 and has wavelength $0.7 \ \mu$m. Show that the magnetic force is indeed small.

6.4   Calculate the lowest frequency of a standing wave in a cavity $20 \ \text{cm} \times 18 \ \text{cm} \times 15 \ \text{cm}$, with perfectly conducting walls. Write down the corresponding fields and show they can be considered as a superposition of plane travelling waves, being reflected back and forth from the walls.

6.5   Show that the total field energy of a standing wave in a cubical cavity is constant, equal to $\varepsilon_0(\alpha_x^2 + \alpha_y^2 + \alpha_z^2) L^3/16$, and oscillates harmonically, at frequency 2ω, between being all electric and all magnetic.

7

The electrostatics of conductors

We have so far in this book regarded the sources ρ, $\mathbf{J}$ of the electromagnetic field as given. However, the charges and currents in material media are themselves driven by the fields, so that we need to describe the electrical and magnetic responses of materials to an electromagnetic field. At the atomic level, the Coulomb forces between electrons and atomic nuclei are responsible for their binding into atoms and molecules, and the large scale structure of materials. A description at this level involves the complicated quantum mechanics of the constituent particles of the materials, and is the province of condensed matter physics and material science. For the most part, we shall rather be concerned with the *macroscopic* electrical and magnetic properties of materials, which can often be described phenomenologically by a small number of parameters, such as the electrical conductivity. These parameters are usually obtained by direct experiment on the material concerned.

We begin in this chapter with a description of conductors in electrostatic equilibrium.

7.1 Electrostatic equilibrium

A *conductor* is a material in which there are electrons, or ions, free to migrate and transport charge in response to an electric field. In a metal or semiconductor the charge carriers are electrons. The principal property of any homogeneous conductor in electrostatic equilibrium is that the electric field $\mathbf{E}(\mathbf{r}) = 0$ at all interior points $\mathbf{r}$. If $\mathbf{E}(\mathbf{r})$ were not zero, the mobile charge carriers would move in response to the mean Coulomb force, until a charge distribution was established for which the condition held. Any attempt to establish an electric field inside a conductor, for example by bringing up an electrically charged body nearby, leads to a redistribution of charge to cancel the imposed field. In an *insulator* the electrons are tightly bound to the ions and there is no significant flow of current in response to an external field; thus external electric fields can

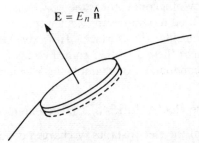

Fig. 7.1 A flat cylinder enclosing surface charge $\sigma\delta S$. Inside the conductor the field, and hence the flux, is zero, and the flux through the sides of the cylinder is also zero. Hence the only flux is $E_n\,\delta S$ through the outer face and, by Gauss's theorem, this equals $\sigma\delta S/\varepsilon_0$.

penetrate an insulator. Insulating materials will be discussed in Chapter 10.

Since, in a conductor in electrostatic equilibrium, $E = 0$ at an interior point, it follows from the Maxwell equation (5.7a), $\nabla \cdot E = \rho/\varepsilon_0$, that $\rho(r) = 0$ also at an interior point. Any positive or negative macroscopic charge density on a conductor must lie at the surface. In the case of a metallic conductor, atomic scale calculations show that this 'surface charge' will be confined to within a few atomic layers of the surface. We can therefore construct an averaged macroscopic *surface charge density* $\sigma(r)$ such that $\sigma(r)\,\delta S$ is the total charge associated with an element of area δS of the surface; the total charge carried by the conductor will be $\int \sigma\,dS$.

The tangential component of the electric field at the surface between two materials is always continuous, even in time-dependent situations. This follows from the integral form (5.4) of the Maxwell equation (5.6b), taking a contour lying on each side of the surface. (See Appendix C and Fig. C.2.) Applying this result to the surface of a conductor in electrostatic equilibrium, we have $E = 0$ inside the conductor, so that the tangential component of the field outside must be zero at the surface.

Writing $E = E_n \hat{n}$ for the field at a point just outside the conductor, an application of Gauss's theorem to a flat cylinder of area δS enclosing surface charge $\sigma\delta S$ (Fig. 7.1) shows that

$$E_n(r) = \sigma(r)/\varepsilon_0. \tag{7.1}$$

Since field lines are perpendicular to the equipotential surfaces, just outside the conductor the potential will have a constant value Φ_{ext}, which is referred to as the *potential of the conductor*. We often put $\Phi_{ext} = V$.

Inside the conductor, since $E = -\nabla\Phi = 0$, the mean potential, Φ_{int} say, is also constant. We note here that a difference exists between Φ_{ext} and Φ_{int}, which depends rather sensitively on the details of the atomic charge distribution at the surface (Problem 7.6).

A related property of a conductor is the *work function* W, which is the minimum work required to remove an electron from inside the conductor to a point just outside the surface. However, W is not simply $(-e)(\Phi_{ext} - \Phi_{int})$, since there are other contributions to the energy of the electron inside the conductor (for example, its kinetic energy).

7.2 The electrostatics of charged conductors

A central problem of the electrostatics of charged conductors is: given a system of conductors in a vacuum, each either carrying a given charge, or held at a given potential, what is the resulting electric field around the conductors?

The solution to this problem will be achieved if we can find the electrostatic potential $\Phi(\mathbf{r})$ which satisfies both Laplace's equation (2.21), $\nabla^2 \Phi = 0$, in the region outside the conductors, and the appropriate boundary conditions over the surfaces of the conductors. If the problem is properly specified it can be shown that there is one and only one solution to it. This result is known as the *uniqueness theorem* (Appendix B).

We shall consider only a few special, but important, problems for which a solution can be found rather easily. In problems with high symmetry it may be possible to use Gauss's theorem in its integral form (2.18) directly. For example, the field due to an isolated conducting sphere of radius a carrying a charge Q must have spherical symmetry and so be of the form $\mathbf{E}(\mathbf{r}) = E(r)\hat{\mathbf{r}}$, where $\mathbf{r}$ is the radial vector from the centre of the sphere. We apply Gauss's theorem to a concentric spherical surface of radius $r > a$. The total flux through this surface is $\int \mathbf{E} \cdot d\mathbf{S} = 4\pi r^2 E(r)$, since for a sphere $d\mathbf{S} = \hat{\mathbf{r}} \, dS$. Hence $4\pi r^2 E(r) = \int \rho \, dV / \varepsilon_0 = Q / \varepsilon_0$, and $E(r) = Q/4\pi\varepsilon_0 r^2$: the field outside the sphere is the same as that which would be produced by a point charge Q at its centre, as also is the potential $\Phi(r) = Q/4\pi\varepsilon_0 r$. Another example of the direct use of Gauss's theorem is given in Problem 7.1.

7.3 The method of images

The 'method of images' relies on an ingenious application of the uniqueness theorem. The simplest case for which it may be used is that of a point charge Q lying at a distance d from a plane surface of a conductor, of extent $\gg d$, which is held at zero potential.

Let us take the conducting surface as the xy-plane, with charge Q at $\mathbf{d} = (0, 0, d)$. Now consider the potential due to a charge Q at $\mathbf{d}$ and its 'image charge' $-Q$ at $-\mathbf{d}$,

$$\Phi(\mathbf{r}) = \frac{Q}{4\pi\varepsilon_0 |\mathbf{r} - \mathbf{d}|} - \frac{Q}{4\pi\varepsilon_0 |\mathbf{r} + \mathbf{d}|}. \tag{7.2}$$

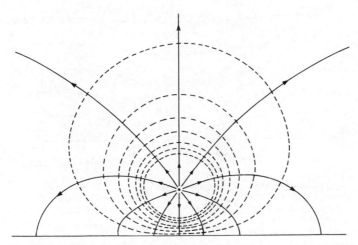

Fig. 7.2 Field lines (with arrows) and sections of some equipotential surfaces (at constant intervals) for a positive point charge at a short distance from a plane conducting surface. Closer to the charge, the spacing of equipotentials with the same interval becomes too close to plot conveniently.

As is immediately evident, the xy-plane is the equipotential surface $\Phi = 0$ and the field lines meeting this plane are, of course, perpendicular to it (Fig. 7.2). In the region $z > 0$, this potential satisfies all the conditions required of the potential for our original problem: it behaves correctly in the neighbourhood of the charge Q, it takes the value zero at the conducting surface $z = 0$, and goes to zero as $r \to \infty$. Thus (7.2) is the required – and unique – solution at points $\mathbf{r}$ where $z > 0$. At points where $z < 0$ in the conductor, the potential is constant; the expression (7.2) is invalid inside the conductor.

In practice, to maintain a conductor at constant potential it is 'grounded', that is to say the conductor is connected by a thin wire to a large body, called 'ground', sufficiently large for its potential to be effectively unchanged if the charge on the conductor to which it is connected changes in response to an external field. Thus 'induced charge' may flow between conductor and ground.

In our example above the induced charge density on the conductor, given by equation (7.1), at a point distance ρ from the origin (see Fig. 7.2) is

$$\sigma(\rho) = \varepsilon_0 E_n = -\frac{Q}{2\pi} \frac{d}{(d^2 + \rho^2)^{\frac{3}{2}}}, \tag{7.3}$$

where E_n is just the resultant of the Coulomb fields of the charge and its image. This induced charge produces an electric field inside the conductor which just cancels that due to the charge Q.

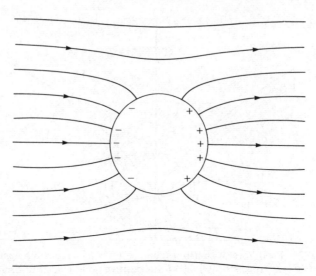

Fig. 7.3 Field lines around a conducting sphere placed in an imposed external field $\mathbf{E}_0$. The distribution of charge on the sphere is also indicated.

The induced charge on the conductor exerts a force on the charge Q which is clearly the same as that calculated from the fictional 'image charge' $-Q$ at $-\mathbf{d}$, so that whatever the sign of Q it is *attracted* to the plate by a force $Q^2/4\pi\varepsilon_0(2d)^2$. This 'image force' is part of the explanation of why excess charge can be held on a conductor, and contributes to the work function. A point charge Q at a distance z (greater than a few ångströms) from a conductor is thus subject to an attractive potential $-Q/(16\pi\varepsilon_0 z)$.

7.4 A conducting sphere in a uniform field

The problem of an isolated uncharged conducting sphere of radius a placed in a much larger region, where the electric field was, previously, uniform, illustrates another style of solution of problems in electrostatics. The total field resulting from the imposed external field $\mathbf{E}_0$ and the induced field clearly has cylindrical symmetry about the axis of the sphere parallel to the field (Fig. 7.3). 't is natural to use spherical polar coordinates (r, θ, ϕ) with the centre of the sphere as origin, and the z-axis parallel to the direction of the imposed field, which then is given by the potential

$$\Phi_0 = -E_0 z = -E_0 r \cos\theta$$

(cf. equation (2.20)); we have taken the potential to be zero at $r = 0$, but this choice is arbitrary.

The effect of the external field on the conducting sphere is to induce a distribution of surface charge, which must at large distances give in leading order a dipole potential. (If the sphere is isolated, it can acquire no net charge.) Since the symmetry of the problem requires that the induced dipole moment lie along Oz, this must be of the form $B\cos\theta/r^2$ (cf. equation (3.1)). Hence the total potential at distances $r \gg a$ must be of the form

$$\Phi(r,\theta) = -E_0 r\cos\theta + \frac{B\cos\theta}{r^2}.$$

It is surprising at first, but true, that a potential of this form provides an exact solution for all $r > a$; if we take $B/a^2 = E_0 a$, so that

$$\Phi(r,\theta) = -E_0\left(r - \frac{a^3}{r^2}\right)\cos\theta, \tag{7.4}$$

the surface of the sphere, $r = a$, is the equipotential surface $\Phi = 0$. We again appeal to the uniqueness theorem: having found a solution which satisfies the boundary conditions of the problem, we know it is the correct solution. The calculation of the redistribution of charge on the sphere is left as an exercise (Problem 7.3).

The expression (7.4) is in fact a special case of the general expansion of the solution of Laplace's equation in spherical harmonics,

$$\Phi(r,\theta,\phi) = \sum_{l,m}\left(A_{lm}r^l + \frac{B_{lm}}{r^{l+1}}\right)Y_{lm}(\theta,\phi),$$

with all terms zero except for A_{10} and B_{10}. You will need to refer to more advanced texts for the development of this theme.

7.5 Capacitance

We have seen that conductors are able to hold charge. This is an important technical property. Let us consider a system of conductors at fixed positions in a vacuum, of which all but one are held at zero potential. Suppose a charge Q_1 placed on the isolated conductor brings it to a potential V_1. Provided that the electrostatic forces cause only negligible mechanical distortions of the conductors, it follows from the linearity of both Laplace's equation and the boundary conditions on the potential, that a charge $Q = \alpha Q_1$ on the conductor would raise it to potential $V = \alpha V_1$. Thus the ratio Q/V is a constant, depending only on the geometry of the system of conductors. We write $Q/V = C$, and call C the *capacitance* of the device. The SI unit of capacitance is the *farad* (F). For example, an isolated sphere of radius a carrying charge Q has potential $V = Q/(4\pi\varepsilon_0 a)$; its capacitance is $C = Q/V = 4\pi\varepsilon_0 a$. If $a = 1$ cm, $C = 1.1 \times 10^{-12}$ F. The farad is in fact an inconveniently large unit, and capacitances are usually quoted in microfarads ($1\ \mu F = 10^{-6}$ F) or picofarads (1 pF $= 10^{-12}$ F).

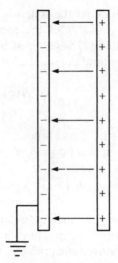

Fig. 7.4 Charge and field lines for a parallel plate capacitor, neglecting
edge effects ($V > 0$).

7.6 The parallel plate capacitor

Technical devices designed to store charge are called *capacitors*. A simple
example is the parallel plate capacitor. This consists of two plane
conducting plates, each of area A say, separated by a distance d (Fig. 7.4).
If $d \ll$ (shortest linear dimension of the plates), edge effects may be
neglected and the potential problem is solved to a good approximation in
the region between the plates by

$$\Phi = (V/d)z,$$

taking the plate held at $\Phi = 0$ to have the surface plane $z = 0$, and the
other plate at potential $\Phi = V$ to have the surface $z = d$. Clearly this
potential satisfies the boundary conditions at the plates, and Laplace's
equation $\nabla^2\Phi = 0$ in the region between them.

The electric field between the plates is

$$\mathbf{E} = -\nabla\Phi = (0, 0, -V/d),$$

so that the surface charge density on the plate at potential V is $\sigma =
\varepsilon_0 E_n = -\varepsilon_0 E_z = \varepsilon_0 V/d$ (and the induced charge density on the grounded
plate is $-\varepsilon_0 V/d$). Thus the total charge on the plate at potential V,
neglecting edge effects, is $Q = \sigma A = \varepsilon_0 VA/d$, and the capacitance is $C =
Q/V = \varepsilon_0 A/d$. The capacitance can be made large by making d small.

As we shall see in Chapter 10, the capacitance can be further increased
by introducing insulating material between the plates. A solid insulator
between the plates will also serve to hold them from collapsing; it is clear
that when charged the plates will experience electrostatic attraction.

7.7 The energy stored in a capacitor

Consider a capacitor, of capacitance C, carrying charge Q and hence at potential $V = Q/C$. To change the charge on it by a small amount dQ, we must transfer charge between the plates. For example, to remove an electron from the surface of the grounded conductor to a point just outside the conductor where the potential is $V_1 = 0$ requires work W_1, where W_1 is the work function of the conductor. Then moving the electron to a point just outside the other plate of the capacitor, where the potential is $V_2 = V$, requires work $(-e)(V_2 - V_1) = -eV$, which is independent of the particular path taken (§2.3). Work W_2 is released when the electron is attached to the surface of the plate. Thus the total work performed on the electron is

$$W_1 - eV - W_2 = -eV + (W_1 - W_2).$$

For simplicity we shall suppose the capacitor plates to be of the same material, so that $W_1 = W_2$; effects which arise when conductors have different work functions will be considered in Chapter 15. Then the work done is just $(-e)V = V\,dQ$. (If an electron is transferred in the other direction, $dQ = +e$; the work done is again $V\,dQ$.) Thus the total work done in charging the capacitor is

$$U = \int_0^Q V\,dQ = \int_0^Q \frac{Q}{C}\,dQ = \frac{Q^2}{2C} = \frac{QV}{2} = \frac{CV^2}{2}. \tag{7.5}$$

In equation (7.5) U is defined relative to the uncharged state of the capacitor, and clearly energies associated with fields at the surfaces are not included. Changes in these are generally negligible if the capacitor is not overloaded.

In the case of capacitors which contain no insulating material, we may express U in terms of the electric field. We have

$$U = \tfrac{1}{2}QV = \frac{1}{2}\int_S \Phi\sigma\,dS, \tag{7.6}$$

where the integral is taken over the surfaces of the capacitor plates and also, if it is necessary, over a surface at 'infinity' enclosing the system, where the potential field $\Phi(\mathbf{r})$ goes to zero. Since by (7.1) the charge density $\sigma = \varepsilon_0 E_n = \varepsilon_0\,\mathbf{E}\cdot\hat{\mathbf{n}}$, we can write (7.6) as

$$U = \frac{\varepsilon_0}{2}\int_S \Phi\mathbf{E}\cdot d\mathbf{S}, \tag{7.7}$$

and then use the divergence theorem to transform this expression into the volume integral

$$U = -\frac{\varepsilon_0}{2}\int_V \boldsymbol{\nabla}\cdot(\Phi\mathbf{E})\,dV.$$

The change of sign occurs since the sense of $d\mathbf{S}$ in (7.7) is outward from the conductors, whereas the convention of the divergence theorem requires it to be outward from the volume of integration.

Using the results $\nabla \cdot (\Phi\mathbf{E}) = \mathbf{E} \cdot \nabla\Phi + \Phi\nabla \cdot \mathbf{E}$, $\mathbf{E} = -\nabla\Phi$, and $\nabla \cdot \mathbf{E} = 0$ in empty space, we are left with

$$U = \frac{\varepsilon_0}{2} \int \mathbf{E}^2 \, dV. \tag{7.8}$$

It should be noted that the field $\mathbf{E}$ in (7.8) is everywhere finite. The energy U is relative to the energy of the uncharged capacitor, and again excludes any surface field effects.

7.8 Forces on charged surfaces

If we apply the result (7.8) to the parallel plate capacitor discussed above, in which the field is $\mathbf{E} = (0, 0, -\sigma/\varepsilon_0)$ between the plates, we find immediately

$$U = \frac{\sigma^2 A d}{2\varepsilon_0}. \tag{7.9}$$

A is the area of the plates, d the distance between them. The force on the plate at $z = d$ is clearly normal to the plate and, keeping the charge constant, is by energy conservation

$$F_z = -\frac{\partial U}{\partial d} = -\frac{\sigma^2 A}{2\varepsilon_0}. \tag{7.10}$$

Thus there is an attractive force between the plates of magnitude $\sigma^2/2\varepsilon_0$ per unit area. The force on an element of area $d\mathbf{S}$ of the surface of either plate may be written as $\sigma^2 \, d\mathbf{S}/2\varepsilon_0$. This expression is generally valid for the force on a charged element of any (smooth) shape of conductor.

Problems

7.1   A wire of radius a carrying charge Q runs along the axis of a long cylindrical conductor of inner radius b and length l which is grounded. Neglect end effects, and use Gauss's theorem to find the electric field, and hence the potential, in the region between the conductors. Show that the capacitance of the system is $2\pi\varepsilon_0 l/\ln(b/a)$.

7.2   Verify by integration of (7.3) that a total charge $-Q$ is induced on the planar surface.

7.3   Show that the surface charge density on the conducting sphere considered in §7.4 is given by

$$\sigma(\theta) = 3\varepsilon_0 E_0 \cos\theta,$$

and the induced moment on the sphere is

$$\mathbf{p} = (4\pi\varepsilon_0 a^3) \mathbf{E}_0.$$

7.4   How is the potential (7.4) modified if the sphere carries charge Q?

7.5   Find an approximate expression for the capacitance per unit length of two long parallel wires, each of radius a, distance d apart, valid for $d \gg a$.

7.6   At the surface of a material, the electron density profile extends further out from the surface than the corresponding positive charge profile, so that, averaging over many surface atoms, there is an effective electric dipole sheet at the surface. As a crude model, suppose surface electrons uniformly distributed on a sheet with number density $0.1\ \text{Å}^{-2}$ are separated from a positive sheet by a distance of $0.5\ \text{Å}$. Calculate the potential drop in passing out of the surface.

7.7   Using cylindrical polar coordinates (ρ, ϕ, z), some special solutions of Laplace's equation are $\Phi = \ln \rho$, $\Phi = \rho \cos \phi$, $\Phi = \cos \phi / \rho$. Verify this (e.g., using Cartesian coordinates).

       Show that the potential $\Phi = - E_0 \rho \cos \phi$ corresponds to a uniform electric field in the x-direction.

       A long straight conducting wire of radius a carrying charge q per unit length is inserted into a region of uniform electric field, and perpendicular to it. Show that the potential is then given by

$$\Phi = - E_0(\rho - a^2/\rho) \cos \phi - (q/2\pi\varepsilon_0) \ln \rho + (\text{constant}).$$

Find an expression for the minimum charge per unit length, $q_{\min}$, for the electric field to be directed out of the wire over all its surface. A long distance down the field, all the field lines leaving the wire are parallel. What is the width of this band of field lines when $q = q_{\min}$?

7.8   Using the expression $\sigma^2 \, dS/2\varepsilon_0$ for the force on a surface element dS, verify that the force on the conducting plate in §7.3 is $Q^2/4\pi\varepsilon_0(2d)^2$.

8

Steady currents in conductors

In a conductor, the mobile charge carriers are set into motion if they experience a non-vanishing mean field $\mathbf{E}$. As we showed in Chapter 1, a macroscopic electric current $\mathbf{J}$ is associated with this motion. The previous chapter was concerned with the equilibrium situation when, inside the conductor, any applied electric field is cancelled by the field of redistributed charge carriers, so that both $\mathbf{E} = 0$ and $\mathbf{J} = 0$ everywhere inside. We now consider what happens when a non-vanishing $\mathbf{E}$ field is maintained inside a conductor, for example by making it part of a circuit in which there is a source of e.m.f.

We shall exclude from the discussion the case of superconductors, which are the subject of Chapter 14, and specially engineered semiconductors, some of whose important properties are discussed in Chapter 15. Here we shall consider the important case of simple conducting solids, in which the charge carriers are electrons.

8.1 Time-independent fields in conductors

It is found experimentally that if a time-independent field $\mathbf{E}(\mathbf{r})$ exists inside a homogeneous isotropic normal conductor, the resulting current density $\mathbf{J}(\mathbf{r})$ is given to a good approximation by the linear relation

▶ $$\mathbf{J}(\mathbf{r}) = \sigma \mathbf{E}(\mathbf{r}). \qquad (8.1)$$

The constant of proportionality σ is called the *conductivity* of the material (Table 8.1). Its inverse, σ^{-1}, is called the *resistivity*. The equation holds in this simple form for sufficiently small fields $\mathbf{E}$. The resistivity depends on the parameters, such as pressure, which determine the state of the material; in particular it may be strongly dependent on the temperature (Fig. 8.1). It is also in principle dependent on the magnetic field $\mathbf{B}$, but in this chapter we shall suppose that all magnetic fields are such that they have negligible effect on the relation between $\mathbf{J}$ and $\mathbf{E}$.

64

Table 8.1. *Electrical conductivities in* $\Omega^{-1}\,m^{-1}$

Material	Conductivity at 0 °C	Classification
Copper	6.45×10^7	
Aluminium	4.00×10^7	Conductors
Iron	1.12×10^7	
Bismuth	0.93×10^6	
Porcelain	$\sim 10^{-10}\text{--}10^{-12}$	
Glass (Pyrex)	$\sim 10^{-12}$	Insulators
Polythene	$\sim 10^{-14}\text{--}10^{-15}$	

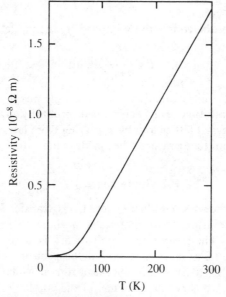

Fig. 8.1 The 'ideal' resistivity of copper as a function of temperature. To this must be added a small sample-dependent 'residual resistivity' which is independent of temperature. (Data from Meaden, G. T. (1966), *Electrical Resistance of Metals*, London: Heywood.)

The linear relation (8.1) can be understood qualitatively from a simple model due to Drude. (This work, published in 1900, followed soon after the discovery of the electron by J. J. Thomson in 1897.) In Drude's model it is supposed that the motion of the conduction electrons is continually being randomised by instantaneous collisions of some sort. Between collisions the electrons are accelerated by the mean field **E** so that

$$m_e \frac{d\mathbf{v}}{dt} = (-e)\,\mathbf{E}.$$

At any instant, say $t = 0$, an electron which had its last collision at time $t = -T$ will have acquired a velocity $v = (-e)ET/m_e$ in the direction of the field E, in addition to its initial random motion.

For the mean value of v at $t = 0$, averaged over all the electrons, we write

$$\bar{v} = (-e)E\tau/m, \qquad (8.2)$$

where τ is called the *collision time*; τ is the mean value of T, averaged over all the electrons.

If there are N conduction electrons per unit volume, the current is (cf. equation (1.13))

$$J = N(-e)\bar{v} = \frac{Ne^2\tau}{m_e}E, \qquad (8.3)$$

and we can identify the conductivity implied by the model to be

▶
$$\sigma = \frac{Ne^2\tau}{m_e}. \qquad (8.4)$$

A quantum mechanical calculation for a simple metal gives a similar result. The collision time τ is determined primarily by the thermal vibrations of the crystal lattice and by scattering from lattice defects. For most metals at room temperature τ lies in the range 10^{-15} s–10^{-14} s.

8.2 Joule heating

In the collision processes, conduction electrons transfer kinetic energy acquired between collisions to the material as a whole; this energy appears as the 'Joule heat'. The mean rate at which work is done on an electron by the field E is $(-e)E \cdot v$. Averaging over all the electrons, the mean value of v is related to J by (8.3). Hence the mean rate at which the field does work on the N electrons per unit volume of the conductor is

$$\text{rate of Joule heating per unit volume} = N(-e)E \cdot \bar{v} = E \cdot J = \sigma E^2 = \frac{J^2}{\sigma}. \qquad (8.5)$$

We have derived these results using the Drude model, but the expression $E \cdot J$ for the Joule heating per unit volume in a conductor is valid generally.

8.3 The flow of current in wires

When the ends of a conducting wire are connected to the terminals of a battery a steady current flow is established in the wire. Since no current flows out through the surface of a wire, we can anticipate that in a thin wire of uniform cross-section S the current density J will be parallel to the axis of the wire, and of constant magnitude. This will be approximately true even around bends in the wire provided the bends are gentle over distances comparable with the diameter of the wire.

From the equation $\mathbf{J} = \sigma\mathbf{E}$, the E field is then also parallel to the axis of the wire and of constant magnitude. The integral $\int \mathbf{E} \cdot d\mathbf{l}$ along any path joining the battery terminals is the terminal potential difference V, so that

$$V = \int_{\text{wire}} \mathbf{E} \cdot d\mathbf{l} = |\mathbf{E}|\, l = |\mathbf{J}|\, l/\sigma, \tag{8.6}$$

where l is the length of the wire (since along the wire E and dl are parallel and $|\mathbf{E}|$ is constant). The total current I flowing in the wire is given by $I = |\mathbf{J}|\, S$, and we can re-write (8.6) in the familiar form of *Ohm's law*:

$$V = IR, \tag{8.7}$$

where

$$R = \frac{l}{\sigma S} \tag{8.8}$$

is the *resistance* of the wire. The SI unit of resistance is the *ohm* (Ω), and hence conductivities are expressed in $\Omega^{-1}\,\text{m}^{-1}$. The linear relation $\mathbf{J} = \sigma\mathbf{E}$ is confirmed by the experimental verification of (8.7) and (8.8).

Since the volume of the wire is Sl, the total rate of Joule heating in the wire is

$$\sigma E^2 Sl = \frac{\sigma V^2 S}{l} = \frac{V^2}{R} = I^2 R = IV, \tag{8.9}$$

where we have used (8.5) and the result $|\mathbf{E}| = V/l$. We have recovered the well-known elementary expressions!

8.4 Steady current flows: the general case

In a conductor in a steady state, Maxwell's equations give $\nabla \times \mathbf{E} = 0$, and so an electric potential Φ exists such that $\mathbf{E} = -\nabla\Phi$ (as we assumed above). Also the continuity equation (1.15) reduces to $\nabla \cdot \mathbf{J} = 0$ when $\partial\rho/\partial t = 0$. Hence in a homogeneous isotropic conductor in which $\mathbf{J} = \sigma\mathbf{E}$, we have $\nabla \cdot \mathbf{E} = 0$, and Φ satisfies Laplace's equation $\nabla^2\Phi = 0$. However the boundary conditions which Φ must satisfy are different from the boundary conditions which apply to this equation in electrostatics. We may suppose that the potentials of the terminals in contact with the conductor which drive the current flow are specified. There can be no current flow normal to the surface of the conductor. Since $\mathbf{J} = \sigma\mathbf{E}$, it follows that the normal component of the electric field, $\mathbf{E} \cdot \mathbf{n} = -\partial\Phi/\partial n$, is zero at the surface of the conductor. These conditions suffice to make the solution to the potential problem unique; if the potential problem is solved, the current flow may be calculated using $\mathbf{J} = \sigma\mathbf{E} = -\sigma\nabla\Phi$.

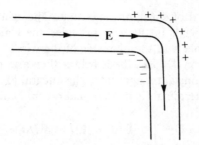

Fig. 8.2 How does a current turn a corner? The field in a wire is established by the surface distribution of charge on the wire.

As a simple example, relevant to the 'leakage current' in a coaxial cable, consider the current flow between two long concentric conducting cylinders, the inner of radius a being held at potential V and the outer of radius b at ground potential. The space between the cylinders is filled with material of conductivity σ.

Neglecting end effects, the solution of $\nabla^2\Phi = 0$ with cylindrical symmetry is of the form (cf. Problem 2.5)

$$\Phi(\rho) = A \ln \rho + B$$

where $\rho = (x^2 + y^2)^{\frac{1}{2}}$, and the boundary conditions at $\rho = a$ and $\rho = b$ give

$$\Phi(\rho) = \frac{V}{\ln(a/b)} \ln(\rho/b).$$

Then

$$J(\rho) = \sigma E(\rho) = -\sigma \frac{\partial\Phi}{\partial\rho} = -\sigma \frac{V}{\ln(a/b)} \frac{1}{\rho}.$$

The total current $I = 2\pi\rho J(\rho) l$, where l is the length of the cable, and we may define the resistance

$$R = \frac{V}{I} = \frac{1}{2\pi\sigma l} \ln(b/a).$$

8.5 Digressions

It is interesting to contemplate just how an electric field of uniform strength can be guided from one terminal of a battery, by a more or less arbitrary path, to the other terminal. Under steady state conditions, it follows from Maxwell's equations that the electric field is determined by the distribution of charges in the system. Hence when a current is flowing in a wire, there must be electric charge distributed on the surface of the wire in such a way as to establish the E field in and around the wire (Fig. 8.2).

The energy dissipated in Joule heating in a wire can be regarded as

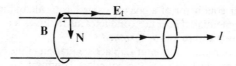

Fig. 8.3 Fields at the surface of a cylindrical wire carrying a steady current.

transported from the battery through the electromagnetic field which surrounds the battery and the wire. We can illustrate this rather simply in the case of a long straight cylindrical wire of radius a, carrying a current I. The magnetic field just outside the wire is given by (4.5), with $\rho = a$. The tangential component of $\mathbf{E}$ is continuous at the surface of the wire (Appendix C) and hence equal to the magnitude E of the field inside the wire. (There is also, in general, a normal component of $\mathbf{E}$ outside the wire due to surface charge, but this does not contribute to the flow of energy into the wire.) The rate of flow of energy into the wire is given by the integral $-\int \mathbf{N} \cdot d\mathbf{S}$ over the surface of the wire, where $\mathbf{N} = \mu_0^{-1}\mathbf{E} \times \mathbf{B}$ is the Poynting vector. The normal component of $\mathbf{N}$ at a point on the surface of the wire is (Fig. 8.3)

$$\mu_0^{-1}|\mathbf{E}_t||\mathbf{B}| = \mu_0^{-1}E(\mu_0 I/2\pi a) = EI/2\pi a,$$

so that the total energy input into a length l of the wire, which has a surface area $2\pi a l$, is EIl. Since $El = V$, where V is the potential difference between the ends of the wire, this is just the rate of Joule heating, IV, which we calculated in §8.3. In a steady state the energy of the electromagnetic field outside the wire remains constant. There must therefore be an equal flow of energy from the battery into the field, through a surface enclosing the battery. Pursuing the energy source into the battery, we would find it in the chemistry of the electrolytic reactions.

Problems

8.1 The conductivity of sodium at 0 °C is $\sigma = 23.8 \times 10^6 \Omega^{-1}\,\mathrm{m}^{-1}$ and the density of conduction electrons is $2.65 \times 10^{28}\,\mathrm{m}^{-3}$. Estimate the collision time τ. A conduction electron in sodium has a kinetic energy of 3.24 eV. How far will it travel in time τ?

8.2 Calculate the resistance of a 1 m length of wire of radius 0.5 mm if it is made of (i) aluminium, (ii) bismuth. Suppose the wires to be joined end to end, and to carry a current of 1 A from the bismuth to the aluminium. What are the electric field strengths in the two wires? What must be the net electric charge at the junction? How many electrons does this represent?

8.3 Two fine wires of radius a are inserted vertically to the same depth l into a large bath of electrolyte of conductivity σ, at a distance $d \gg a$ apart. Assuming the resistance of the wires to be negligible, find an approximate expression for the resistance per unit length of the wires to a current passing between them when $l \gg d$ (cf. Problem 7.5).

8.4   The outer conductor of a coaxial cable has inside radius b and is held at ground potential. The inner conductor of radius a is at ground potential where $z = 0$ and carries a current I.

Show that the solution of Laplace's equation in the region between the conductors which satisfies the boundary conditions is

$$\Phi(\rho, z) = -\frac{Iz}{\sigma\pi a^2}\frac{\ln(\rho/b)}{\ln(a/b)}$$

where $\rho = (x^2 + y^2)^{\frac{1}{2}}$, and σ is the conductivity of the conductor.

Hence show that the wire carries a surface charge density equal to $\varepsilon_0 Iz/\sigma\pi a^3 \ln(a/b)$.

If $a = 1$ mm, $b = 10$ mm, $I = 1$ A, and the conductor is of copper, show that, at $z = 1$ m, this corresponds to an excess of about 10^6 electrons per metre of conductor.

9

Magnetostatics

In this chapter we obtain expressions for the magnetic field $\mathbf{B(r)}$ produced by steady currents flowing in conductors, when no magnetic materials are present. You will probably have seen, and used, these expressions before. We shall show that they are not merely *ad hoc* rules, but that they follow from Maxwell's equations.

In a steady state the currents and fields are constant in time, so that the Maxwell equations (5.7b) and (5.7c) reduce to

$$\nabla \times \mathbf{B} = \mu_0 \mathbf{J}, \qquad (9.1\text{a})$$

$$\nabla \cdot \mathbf{B} = 0, \qquad (9.1\text{b})$$

or, in integral form,

$$\int_\Gamma \mathbf{B} \cdot d\mathbf{l} = \mu_0 \int_{S(\Gamma)} \mathbf{J} \cdot d\mathbf{S}, \qquad (9.2\text{a})$$

$$\int_S \mathbf{B} \cdot d\mathbf{S} = 0. \qquad (9.2\text{b})$$

The latter pair of equations, Ampère's law and Gauss's theorem for the magnetic field respectively, can be used directly to obtain the magnetic field when the current distribution possesses a high degree of geometrical symmetry: for example the field in a long solenoid (§9.5), or that in a coaxial cable (Problem 4.7). However, we shall obtain more generally applicable results, and shall also use these in the calculation of self and mutual inductances. You are advised to refresh your memory of vector identities, and the theorems set out in the Mathematical Prologue, before continuing with this chapter!

9.1 The vector potential

The solution of equations (9.1) for the field $\mathbf{B}$, given the current distribution $\mathbf{J(r)}$, is very much expedited by the introduction of a new vector field: the *vector potential* $\mathbf{A}(\mathbf{r})$. Since $\nabla \cdot \mathbf{B} = 0$, we can always write $\mathbf{B}$ as the curl of another vector field, say

$$\blacktriangleright \qquad\qquad \mathbf{B} = \nabla \times \mathbf{A}. \qquad\qquad (9.3)$$

This is a mathematical theorem of vector analysis (Problem 9.1). The condition $\nabla \cdot \mathbf{B} = 0$ is satisfied identically, since $\nabla \cdot \nabla \times \mathbf{A} \equiv 0$.

The field $\mathbf{A}(\mathbf{r})$ is *not unique*. If $\mathbf{B} = \nabla \times \mathbf{A}$, then also $\mathbf{B} = \nabla \times \mathbf{A}'$ where

$$\blacktriangleright \qquad\qquad \mathbf{A}'(\mathbf{r}) = \mathbf{A}(\mathbf{r}) + \nabla \chi(\mathbf{r}), \qquad\qquad (9.4)$$

and $\chi(\mathbf{r})$ is an arbitrary function. This result holds because $\nabla \times \nabla \chi \equiv 0$. The vector potentials $\mathbf{A}$ and $\mathbf{A}'$ describe the same magnetic field. The relation (9.4) is called a *gauge transformation*. We can exploit this freedom in the choice of vector potential to simplify the mathematics.

Setting $\mathbf{B} = \nabla \times \mathbf{A}$, equation (9.1a) becomes, by a vector identity,

$$\nabla \times (\nabla \times \mathbf{A}) = \nabla(\nabla \cdot \mathbf{A}) - \nabla^2 \mathbf{A} = \mu_0 \mathbf{J}.$$

We now choose to work in the gauge in which $\nabla \cdot \mathbf{A} = 0$ everywhere (see Problem 9.3), so that the equation for $\mathbf{A}$ reduces to

$$\nabla^2 \mathbf{A} = -\mu_0 \mathbf{J}. \qquad\qquad (9.5)$$

Each component of this equation, say

$$\nabla^2 A_x = -\mu_0 J_x,$$

has a similar form to Poisson's equation (2.19). Just as charge density is the source of the electrostatic potential, so the current density is the source of the vector potential, and we have immediately the solution

$$\blacktriangleright \qquad\qquad \mathbf{A}(\mathbf{r}) = \frac{\mu_0}{4\pi} \int \frac{\mathbf{J}(\mathbf{r}')}{|\mathbf{r} - \mathbf{r}'|} \, dV'. \qquad\qquad (9.6)$$

Also, taking the divergence of equation (9.5),

$$\nabla^2(\nabla \cdot \mathbf{A}) = -\mu_0 \nabla \cdot \mathbf{J}.$$

But in a steady state, $\nabla \cdot \mathbf{J} = 0$ everywhere, by the continuity equation (1.15), so that

$$\nabla^2(\nabla \cdot \mathbf{A}) = 0$$

everywhere. From the uniqueness theorem (Appendix B), if $\nabla \cdot \mathbf{A} = 0$ at infinity then $\nabla \cdot \mathbf{A} = 0$ everywhere. In the case of a bounded distribution of current density, it is clear from (9.6) that $\nabla \cdot \mathbf{A}(\mathbf{r}) \to 0$ as $|\mathbf{r}| \to \infty$; the solution (9.6) therefore satisfies our gauge condition.

From the expression (9.6) for $\mathbf{A}$, we can calculate $\mathbf{B}(\mathbf{r})$:

$$\mathbf{B}(\mathbf{r}) = \nabla \times \mathbf{A}(\mathbf{r}) = \frac{\mu_0}{4\pi} \int \nabla \times \left(\frac{\mathbf{J}(\mathbf{r}')}{|\mathbf{r} - \mathbf{r}'|} \right) dV'$$

$$= \frac{\mu_0}{4\pi} \int \nabla \left(\frac{1}{|\mathbf{r} - \mathbf{r}'|} \right) \times \mathbf{J}(\mathbf{r}') \, dV'$$

(using the identity $\nabla \times (u\mathbf{F}) = \nabla u \times \mathbf{F} + u\nabla \times \mathbf{F}$ and noting ∇ does not act on $\mathbf{r}'$), or

$$\blacktriangleright \qquad \mathbf{B}(\mathbf{r}) = \frac{\mu_0}{4\pi} \int \frac{\mathbf{J}(\mathbf{r}') \times (\mathbf{r} - \mathbf{r}')}{|\mathbf{r} - \mathbf{r}'|^3} \, \mathrm{d}V'. \qquad (9.7)$$

This is the general expression for the magnetic field of a current distribution.

In the case of a current I flowing in a thin wire, we have already in §4.4 made the identification

$$\mathbf{J} \, \mathrm{d}V \to I \, \mathrm{d}\mathbf{l}, \qquad (9.8)$$

which turns (9.7) into

$$\blacktriangleright \qquad \mathbf{B}(\mathbf{r}) = \frac{\mu_0 I}{4\pi} \int \frac{\mathrm{d}\mathbf{l}' \times (\mathbf{r} - \mathbf{r}')}{|\mathbf{r} - \mathbf{r}'|^3}. \qquad (9.9)$$

This result, the *Biot–Savart* law, was established experimentally in 1820. It reduces the calculation of the magnetic field produced by currents flowing in thin wires to the evaluation of line integrals (Problems 9.4, 9.5).

9.2 Magnetic dipoles

In Chapter 3, we showed that the electric field at points far from a localised distribution of charge can be described in terms of the multipole expansion. If the total charge of the distribution is zero, the next term in the expansion is of the electric dipole form (3.5). A similar expansion can be made for the magnetic field a long way from a localised current distribution, except that the magnetic 'charge' is always zero, and the leading term is, usually, of dipole form.

As in §3.1, for $r \gg r'$ we expand

$$\frac{1}{|\mathbf{r} - \mathbf{r}'|} = \frac{1}{r} + \frac{\mathbf{r} \cdot \mathbf{r}'}{r^3} + \dots,$$

to obtain from (9.6)

$$\mathbf{A}(\mathbf{r}) = \frac{\mu_0}{4\pi} \left[\frac{1}{r} \int \mathbf{J}(\mathbf{r}') \, \mathrm{d}V' + \frac{1}{r^3} \int (\mathbf{r} \cdot \mathbf{r}') \, \mathbf{J}(\mathbf{r}') \, V' \dots \right].$$

Now for a time-independent current distribution,

$$\int \mathbf{J}(\mathbf{r}') \, \mathrm{d}V' = 0 \qquad (9.10a)$$

and

$$\int (\mathbf{r} \cdot \mathbf{r}') \, \mathbf{J}(\mathbf{r}') \, \mathrm{d}V' = \frac{1}{2} \int (\mathbf{r}' \times \mathbf{J}(\mathbf{r}')) \times \mathbf{r} \, \mathrm{d}V'. \qquad (9.10b)$$

(A proof of these identities is sketched in Problem 9.6.) Hence to leading order

▶
$$A(r) = \frac{\mu_0}{4\pi} \frac{m \times r}{r^3} \qquad (9.11)$$

where

▶
$$m = \frac{1}{2} \int r' \times J(r') \, dV' \qquad (9.12)$$

is called the *magnetic moment* of the current distribution. Calculating $B = \nabla \times A$ from (9.11) does indeed give a field of dipole form. After some algebra, we find

$$B(r) = \frac{\mu_0}{4\pi r^3} \left[\frac{3(m \cdot r) \, r}{r^2} - m \right]. \qquad (9.13)$$

9.3 Current loops

For a loop of thin wire in which a current I is flowing, (9.12) becomes

$$m = \tfrac{1}{2} I \oint r \times dl. \qquad (9.14)$$

This can also be written as an integral over a surface S spanning the loop:

▶
$$m = I \int dS, \qquad (9.15)$$

by another mathematical identity (Problem 9.7). m is proportional to the total vector surface area; this does not depend on the particular surface chosen.

An infinitesimal current loop has a magnetic moment $\delta m = I \delta S$ and this will give a dipole field at any finite distance. A finite loop carrying a current I is equivalent to a 'net' of infinitesimal loops over a surface S (Fig. 9.1) so that – remarkably – a loop of current produces a magnetic field which is everywhere equivalent by equation (9.15) to that of a uniform magnetic dipole sheet (except at points actually on the sheet).

In a uniform external magnetic field B a magnetic dipole of moment m experiences a torque $m \times B$, and has potential energy $-m \cdot B$; these expressions are analogous to the results (3.9) and (3.10) for an electric dipole in an electric field. (See Problem 9.8.)

9.4 Mutual and self inductance

Inductance is an important concept in circuit theory. A time-varying current in a circuit produces a time-varying magnetic field; the flux is linked to the circuit itself and any neighbouring circuits. From Faraday's

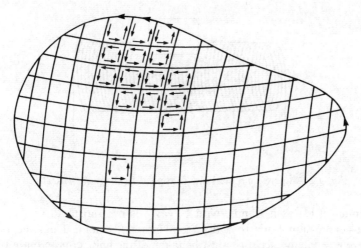

Fig. 9.1 A 'net' of infinitesimal current loops. When the infinitesimal loops are superposed, only the current round the bounding circuit is uncancelled.

law, the time-varying flux induces electromotive forces around the circuits. In elementary circuit theory the *mutual inductance* between a circuit (1) and a circuit (2), L_{12}, is defined to be the magnetic flux through circuit (1) produced per unit current flowing in circuit (2):

$$L_{12} I_2 = \int \mathbf{B}_2 \cdot d\mathbf{S}_1$$

$$= \int \mathbf{\nabla} \times \mathbf{A}_2 \cdot d\mathbf{S}_1, \text{ using the vector potential,}$$

$$= \oint_{\Gamma_1} \mathbf{A}_2 \cdot d\mathbf{l}_1, \text{ by Stokes's theorem.}$$

Using equation (9.6) for $\mathbf{A}_2$ gives

$$L_{12} = \frac{\mu_0}{4\pi} \oint_{\Gamma_1} \oint_{\Gamma_2} \frac{d\mathbf{l}_1 \cdot d\mathbf{l}_2}{|\mathbf{r}_1 - \mathbf{r}_2|}. \tag{9.16}$$

This expression – Neumann's formula – depends only on the geometry of the two circuits. It is completely symmetrical in coordinates (1) and (2), giving the important result that

$$L_{12} = L_{21}. \tag{9.17}$$

The *self inductance* L of a circuit is sometimes defined as the flux through that circuit produced per unit current flowing in it. However, applying the

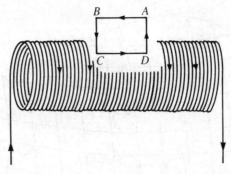

Fig. 9.2 Ampère's circuital law is applied to the path $ABCD$.

formula (9.16) as it stands, with $\Gamma_1 = \Gamma_2$, is not satisfactory: this 'thin wire' expression leads to a divergent integral, coming from the points where $\mathbf{r}_1 = \mathbf{r}_2$, and account must be taken of the finite cross-section of the wire. It is better to define self inductance through the expressions for magnetic energy we shall obtain in §9.6.

The SI unit of inductance is the *henry* (H).

9.5 Solenoids

We illustrate some of the concepts in this chapter with the case of a tightly wound solenoid, carrying a current I. Take cylindrical polar coordinates (ρ, ϕ, z) with the axis of the coil as the z-axis. Since the coil is tightly wound, the field will have axial symmetry, and hence not depend on ϕ, to a good approximation.

Outside the coil the field will be small except near the ends. This can be seen by replacing each turn of the coil by its equivalent dipole sheet: the field *outside* the coil can be thought of as due to the superposition of these sheets, which cancel out except at the ends of the solenoid, where equivalent magnetic 'poles' appear. For a long solenoid these apparent poles are well separated, and the field they produce will be small, except near the ends.

Applying Ampère's circuital law to a path $ABCD$ (Fig. 9.2) away from the ends of the solenoid,

$$\oint_{ABCD} \mathbf{B} \cdot d\mathbf{l} = \mu_0 nI \cdot CD,$$

where n is the number of turns of the coil per unit length. Assume the $\mathbf{B}$ field in this region has no radial component. Since $\mathbf{B}$ is small along AB and $\mathbf{B} \cdot d\mathbf{l} = 0$ along BC and DA, we are left with

$$B_z \cdot CD \approx \mu_0 nI \cdot CD.$$

Hence inside the coil, B_z is constant, independent of ρ, and is given by

$$B_z \approx \mu_0 nI. \tag{9.18}$$

Our assumption that $B_\rho = 0$ is consistent with this: a non-zero B_ρ, independent of ϕ, would violate Gauss's theorem.

Taking circular paths around the axis, Ampère's law gives $B_\phi(\rho) \approx 0$ inside the coil, and $B_\phi(\rho) \approx \mu_0 I/2\pi\rho$ outside the coil (since a current I is flowing down the solenoid).

Thus the field is essentially constant inside the solenoid. A more exact expression for the field at any point on the axis can be found fairly easily using the Biot–Savart law.

The self inductance L of a solenoid can be estimated from the flux linkage. If the solenoid has N turns in all and is of radius a,

$$LI \approx N\pi a^2 B_z = N\pi a^2 \mu_0 nI,$$

giving $L = \mu_0 Nn\pi a^2$. The mutual inductance between coaxial solenoids can be estimated similarly (Problem 9.13).

9.6 Expressions for the magnetic energy

Consider the field $\mathbf{B(r)}$ produced by a steady current distribution $\mathbf{J(r)}$. From §6.5, the total energy stored in the magnetic field is the integral over all space

$$U = \frac{1}{2\mu_0} \int \mathbf{B}^2 \, dV. \tag{9.19}$$

In non-ferromagnetic materials, such as a current carrying wire, the expression is only slightly modified, as is its interpretation. We shall return to this in §11.5.

The expression can be written in other useful forms. Using the vector potential, and a vector identity,

$$U = \frac{1}{2\mu_0} \int \mathbf{B} \cdot \nabla \times \mathbf{A} \, dV$$

$$= \frac{1}{2\mu_0} \int \mathbf{A} \cdot \nabla \times \mathbf{B} \, dV + \frac{1}{2\mu_0} \int \nabla \cdot (\mathbf{A} \times \mathbf{B}) \, dV.$$

The second integral here can be transformed using the divergence theorem into an integral over a sphere of large radius R, and $\to 0$ as $R \to \infty$. Thus, with the help of the Maxwell equation $\nabla \times \mathbf{B} = \mu_0 \mathbf{J}$, we obtain

$$U = \frac{1}{2} \int \mathbf{A} \cdot \mathbf{J} \, dV. \tag{9.20}$$

Using (9.6) for $\mathbf{A}(\mathbf{r})$ gives the symmetrical result

$$U = \frac{1}{2}\frac{\mu_0}{4\pi}\iint \frac{\mathbf{J}(\mathbf{r})\cdot\mathbf{J}(\mathbf{r}')}{|\mathbf{r}-\mathbf{r}'|}\,dV\,dV'. \tag{9.21}$$

The expressions (9.20) and (9.21) should be compared with the expressions (2.24) and (2.25) for the electrostatic energy.

For a set of currents I_i flowing in circuits Γ_i, we can write (9.21) in the form

$$U = \frac{1}{2}\sum_{i,j} I_i I_j \frac{\mu_0}{4\pi}\oint\oint \frac{d\mathbf{l}_i\cdot d\mathbf{l}_j}{|\mathbf{r}_i-\mathbf{r}_j|}$$

$$= \frac{1}{2}\sum_{i,j} L_{ij} I_i I_j. \tag{9.22}$$

In particular, for a single current loop

$$U = \tfrac{1}{2}LI^2. \tag{9.23}$$

These results are useful in the calculation of coefficients of inductance (Problem 9.10).

9.7 Magnetic dipole moments at the atomic level

The magnetic moment of an atom due to the orbital motion of its electrons is given by the atomic form of (9.12):

$$\mathbf{m} = \frac{1}{2}\int \mathbf{r}\times\mathbf{J}_{\mathrm{at}}(\mathbf{r})\,dV,$$

where $\mathbf{J}_{\mathrm{at}}(\mathbf{r})$ is the atomic current. Neglecting the motion of the heavy nucleus, which we take as origin,

$$\mathbf{J}_{\mathrm{at}}(\mathbf{r}) = -e\sum_i \mathbf{v}_i\,\delta(\mathbf{r}-\mathbf{r}_i),$$

where $\mathbf{v}_i$ is the velocity of the ith electron at $\mathbf{r}_i$. Then

$$\mathbf{m} = -\frac{e}{2}\sum_i \int \mathbf{r}\times\mathbf{v}_i\,\delta(\mathbf{r}-\mathbf{r}_i)\,dV$$

$$= -\frac{e}{2}\sum_i \mathbf{r}_i\times\mathbf{v}_i \tag{9.24}$$

$$= -\frac{e\hbar}{2m_\mathrm{e}}\left(\frac{\mathbf{L}}{\hbar}\right)$$

where $\mathbf{L}$ is the total angular momentum of the electrons about the nucleus and we have introduced Planck's constant $\hbar$ into the expression. $(\mathbf{L}/\hbar)$ is dimensionless. The quantity

$$\mu_B = e\hbar/2m_e \approx 9.27 \times 10^{-24} \text{ J T}^{-1}$$

is called the *Bohr magneton*. We know from quantum mechanics that any component of $(\mathbf{L}/\hbar)$, $L_z/\hbar$ say, takes on only integral value $0, \pm 1, \pm 2, \ldots$ so that the magnetic moment of an atom is conveniently measured in units of the Bohr magneton.

It is found experimentally that, in addition to any magnetic moment it may produce due to its orbital motion, an electron has a permanent *intrinsic magnetic moment* of magnitude μ_B, aligned with its intrinsic spin. The intrinsic magnetic moment of the electron is largely responsible for the phenomenon of ferromagnetism.

Protons and neutrons, also, possess permanent magnetic moments, but these are smaller in magnitude by a factor $\sim$ electron mass/nucleon mass.

Problems

9.1 A magnetic field $\mathbf{B(r)}$ is negligibly small at large distances. Construct a corresponding vector potential which has $A_z(\mathbf{r}) = 0$, and $A_x(\mathbf{r})$ and $A_y(\mathbf{r})$ given in terms of integrals of $\mathbf{B(r)}$.

9.2 Devise a vector potential that corresponds to a uniform field $\mathbf{B}_0$ in the z-direction.

9.3 Consider a vector potential $\mathbf{A(r)}$ such that $\nabla \cdot \mathbf{A(r)} = \alpha(\mathbf{r})$ is not identically zero. Show that a gauge transformation $\mathbf{A'(r)} = \mathbf{A(r)} + \nabla\chi(\mathbf{r})$ can be made to give $\nabla \cdot \mathbf{A'(r)} = 0$, and find an expression for $\chi(\mathbf{r})$ in terms of $\alpha(\mathbf{r})$.

9.4 Using the Biot–Savart law, show that the magnetic field at a distance z along the axis of a circular current loop of radius a and carrying a current I is along the axis and of strength $B = \frac{1}{2}I\mu_0 a^2/(a^2 + z^2)^{3/2}$.

A standard way of producing a region of nearly uniform magnetic field is to use two such loops at a distance $2z_0$ apart on their common axis ('Helmholtz coils'). Show that the field on the axis near the midpoint between the coils is highly uniform if $2z_0 = a$.

9.5 A helix of radius a which has its axis along O_z can be parameterised as $\mathbf{r} = (a\cos\phi, a\sin\phi, \alpha\phi)$. If it has n turns per unit length, $\alpha = 1/2\pi n$. A current I flows in such a helical wire. Show that the component of the magnetic field along the axis is $\mu_0 nI$.

$$\left(\text{Note } \int_{-\infty}^{\infty} \frac{dq}{(1+q^2)^{3/2}} = 2. \right)$$

9.6 (a) Show that for any scalar function $F(\mathbf{r})$ and localised time-independent current distribution $\mathbf{J(r)}$,

$$\int \mathbf{J(r')} \cdot \nabla' F(\mathbf{r'}) \, dV' = 0.$$

(b) Taking $F(\mathbf{r'}) = \mathbf{a} \cdot \mathbf{r'}$, where $\mathbf{a}$ is an arbitrary constant vector, show that $\int \mathbf{J(r')} \, dV' = 0$.

(c) Taking $F(\mathbf{r}') = (\mathbf{a}\cdot\mathbf{r}')(\mathbf{r}\cdot\mathbf{r}')$, show that

$$\int (\mathbf{r}\cdot\mathbf{J}(\mathbf{r}'))\,\mathbf{r}'\,\mathrm{d}V' = -\int \mathbf{J}(\mathbf{r}')(\mathbf{r}\cdot\mathbf{r}')\,\mathrm{d}V'.$$

(d) Hence, expanding the triple vector product, prove (9.10b).

9.7   Show that the z-component of the equation

$$\int \mathrm{d}\mathbf{S} = \frac{1}{2}\oint \mathbf{r}\times\mathrm{d}\mathbf{l}$$

is obtained by putting $\mathbf{F} = \frac{1}{2}(-y, x, 0)$ in Stokes's theorem (P.8).

9.8   The torque on a current distribution in an external magnetic field is $\int \mathbf{r}\times(\mathbf{J}\times\mathbf{B})\,\mathrm{d}V$. If $\mathbf{B}$ is uniform, show that this is equal to $\mathbf{m}\times\mathbf{B}$. (Expand the triple vector products. Use Problem 9.6(a) with $F(\mathbf{r}) = \mathbf{r}^2$ to show $\int \mathbf{J}\cdot\mathbf{r}\,\mathrm{d}V = 0$, and Problem 9.6(c).)

9.9   A nucleus has an intrinsic angular momentum $\mathbf{J}$ of constant magnitude, and an intrinsic magnetic dipole moment $\mathbf{m} = \gamma\mathbf{J}$. Show that in an external magnetic field $\mathbf{B}$, the direction of $\mathbf{J}$ precesses about $\mathbf{B}$ with angular frequency γB. (The measurement of this frequency provides a precise way of measuring magnetic field strengths.)

9.10   Show that when a current I flows in the coaxial cable of Problem 4.7, the magnetic energy per unit length in the region between the conductors is $(\mu_0 I/4\pi)\ln(b_1/a)$.
       If $a \ll b_1$, and $(b_2-b_1) \ll b_1$ show that the self inductance per unit length of the cable is approximately $(\mu_0/2\pi)\ln(b_1/a)$.

9.11   Show using equation (9.22) that since $U > 0$, $L_{12} \leqslant \sqrt{(L_{11}L_{22})}$.

9.12   Use Neumann's formula (9.16) to show that the mutual inductance of two concentric circular wire loops lying in a plane, and having radii R_1 and R_2 is

$$M = \frac{\mu_0}{4\pi}2\pi R_2 \int_0^{2\pi} \frac{R_1\cos\theta\,\mathrm{d}\theta}{[(R_1\cos\theta - R_2)^2 + R_1^2\sin^2\theta]^{\frac{1}{2}}},$$

and hence that

$$M = \mu_0(R_1 R_2)^{\frac{1}{2}}G\left(\frac{R_2 - R_1}{(R_1 R_2)^{\frac{1}{2}}}\right)$$

where

$$G(\alpha) = \int_0^\pi \frac{\cos\theta\,\mathrm{d}\theta}{[2(1-\cos\theta)+\alpha^2]^{\frac{1}{2}}}.$$

Show that for large $\alpha, G(\alpha) \approx \pi/2\alpha^3$.
       For small $\alpha, G(\alpha)$ becomes large. Compute $G(\alpha)$ for $\alpha = 0.01, 0.1, 0.5, 1, 2, 3$ in order to draw a graph.

9.13   Two long cylindrical solenoids of radii a and b $(a \lesssim b)$ each have n turns per metre, and one is inserted into the other to a distance z. Show that the mutual inductance is $L_{ab} = L_{ba} \approx \mu_0 n^2\pi a^2 z$.

9.14   A current I_0 is flowing in a circuit of resistance R and self inductance L, when the sustaining e.m.f. is removed. By equating the rate of loss of magnetic energy from the magnetic field to the rate of Joule heating, show that $L\,\mathrm{d}I/\mathrm{d}t + IR = 0$, so that the current falls exponentially as $I = I_0\,\mathrm{e}^{-Rt/L}$.

9.15   Show that the self inductance of a circuit of wire of radius a in the form of a ring of radius b, where $b \gg a$, is given approximately by $L \approx \mu_0 b \ln (b/a)$.

9.16   The flux through a superconducting circuit does not change with time. Two such circuits are rigid, and have fixed orientations in space. The circuits have self inductances L_{11} and L_{22}, and carry currents I_1 and I_2. The mutual inductance $L_{12}(R)$ depends on the position $\mathbf{R}$ of circuit (2) relative to circuit (1). By considering energy changes in a displacement $\delta \mathbf{R}$, show that the force acting on circuit (2) is $I_1 I_2 \nabla_{\mathbf{R}} L_{12}$.

10

Insulators

In contrast to a conductor, an electrical insulator can support a mean electric field in its interior with no flow of current. A perfect insulator has zero conductivity, infinite resistivity. At the atomic level the electrons are not free to flow through an insulator in response to an applied electric field. However, electrons and atomic nuclei are pulled in opposite directions by the field, and respond by a restricted movement that results in a state of *electrical polarisation*, which we describe in this chapter, and which has important consequences for the electrical properties of insulating materials.

10.1 The polarisation field

As a simple model of a solid insulator, consider an assembly of neutral atoms at fixed sites $\mathbf{R}_i$. In the presence of an external electric field, each atom will acquire an electric dipole moment, $\mathbf{p}_i = 4\pi\varepsilon_0\,\alpha\mathbf{E}(\mathbf{R}_i)$, where $4\pi\varepsilon_0\,\alpha$ is its atomic polarisability (§3.1). We showed in Chapter 3 that the dipole moment of a neutral system,

$$\mathbf{p} = \int \mathbf{r}\rho(r)\,\mathrm{d}V, \tag{10.1}$$

does not depend on its position in space, and it follows from this that the dipole moment of a number of distinct neutral atoms is the sum of their individual dipole moments.

Let $N(\mathbf{r})$ be the number density of the atoms in the insulator. Consider a volume δV around the point $\mathbf{r}$. δV is chosen to be small but macroscopic, so that the number of atoms $N(\mathbf{r})\,\delta V$, contained in δV, is large. Then the total dipole moment of the volume element δV is

$$N(\mathbf{r})\,\bar{\mathbf{p}}\delta V,$$

where $\bar{\mathbf{p}}$ is the mean dipole moment of the atoms contained in δV. We define

$$\mathbf{P}(\mathbf{r}) = N(\mathbf{r})\,\bar{\mathbf{p}} \tag{10.2}$$

to be the *polarisation field* of the insulator. If in our model the material is so diffuse that the interaction of one dipole with another can be neglected then

$$\mathbf{P}(\mathbf{r}) = 4\pi\varepsilon_0 \, \alpha N(\mathbf{r}) \, \mathbf{E}(\mathbf{r}). \tag{10.3}$$

The dipole moment of any finite volume V is the sum of the contributions of elements δV and can be written

$$\blacktriangleright \qquad \text{dipole moment} = \sum \mathbf{P}(\mathbf{r}) \, \delta V = \int_V \mathbf{P}(\mathbf{r}) \, \mathrm{d}V. \tag{10.4}$$

We have defined the polarisation field for a simple model, but we can define a $\mathbf{P}(\mathbf{r})$ for *any* insulating medium, even if the underlying microscopic state is not easily described: $\mathbf{P}(\mathbf{r})$ is a smooth macroscopic field such that (10.4) holds for any macroscopic volume V.

There are in fact several physical mechanisms that can contribute to $\mathbf{P}(\mathbf{r})$, in addition to the atomic polarisability of our example. In ionic materials the positive and negative ions are themselves pulled in opposite directions by an external electric field, which results in an ionic contribution to the polarisability. Also, a molecule may have a permanent dipole moment associated with its structure. In the absence of an external electric field, the molecular orientations in gaseous insulators are randomised by collisions, but in an external electric field a molecule with a dipole moment will tend to align with the field, and thereby contribute to the polarisation (Problem 10.2). In condensed matter, more complex phenomena may occur.

10.2 The equivalent charge distribution

To the macroscopic polarisation field $\mathbf{P}(\mathbf{r})$ there must correspond a macroscopic distribution of charge. We can identify this using the mathematical equality

$$\int_V \mathbf{P} \, \mathrm{d}V = \int_V \mathbf{r}(-\boldsymbol{\nabla} . \mathbf{P}) \, \mathrm{d}V + \int_S \mathbf{r}(\mathbf{P} \cdot \mathrm{d}\mathbf{S}), \tag{10.5}$$

where S is the surface enclosing the volume V. This vector equation may be proved by considering the components. For example, it is easy to check by direct differentiation that

$$P_x = -x\boldsymbol{\nabla} \cdot \mathbf{P} + \boldsymbol{\nabla} \cdot (x\mathbf{P}).$$

Integrating over volume, and applying the divergence theorem to the second term on the right, gives the x-component of (10.5).

If we take the surface S to enclose and lie outside the region of insulating material, $\mathbf{P} = 0$ on S and we are left with

$$\text{dipole moment} = \int_V \mathbf{P} \, \mathrm{d}V = \int_V \mathbf{r}(-\boldsymbol{\nabla} \cdot \mathbf{P}) \, \mathrm{d}V. \tag{10.6}$$

This holds for any state of polarisation $\mathbf{P}$. Recalling the definition of dipole moment (10.1):

$$\text{dipole moment} = \int_V \mathbf{r}\rho(\mathbf{r})\,\mathrm{d}V,$$

we can identify $\rho_\mathrm{b} = -\nabla\cdot\mathbf{P}$ as the macroscopic density of 'bound' charge giving the polarisation.

The bound charge must, of course, appear in the macroscopic Maxwell equations, so that at an interior point of an insulator

$$\nabla\cdot\mathbf{E} = \rho_\mathrm{b}/\varepsilon_0 = -\nabla\cdot\mathbf{P}/\varepsilon_0. \tag{10.7}$$

There can also be contributions to ρ_b from higher order induced moments associated with a volume element δV, but these are usually small and we shall neglect them.

We shall find it useful to define the *electric displacement* field $\mathbf{D}(\mathbf{r})$ by

▶ $$\mathbf{D}(\mathbf{r}) = \varepsilon_0\,\mathbf{E}(\mathbf{r}) + \mathbf{P}(\mathbf{r}). \tag{10.8}$$

Then (10.7) becomes $\nabla\cdot\mathbf{D} = 0$.

If there are additional 'external' charges present, the Maxwell equation for $\mathbf{E}$ is

$$\nabla\cdot\mathbf{E} = (\rho_\mathrm{b} + \rho_\mathrm{ext})/\varepsilon_0 = (-\nabla\cdot\mathbf{P} + \rho_\mathrm{ext})/\varepsilon_0, \tag{10.9}$$

so that $\mathbf{D}$ satisfies

▶ $$\nabla\cdot\mathbf{D} = \rho_\mathrm{ext}. \tag{10.10}$$

Thus the only source of $\mathbf{D}$ is the external charge.

So far, the introduction of the $\mathbf{D}$ field has been purely formal. In order to make progress we need to know how $\mathbf{P}$ is related to $\mathbf{E}$ (or to other fields, for example strain fields). Such *constitutive relationships* are inferred from experiment.

10.3 The electric susceptibility and the dielectric constant

In most insulators, whether gases, liquids or solids, the polarisation field is principally determined by the mean electric field, and in most insulators it is found experimentally that a time-independent field $\mathbf{E}(\mathbf{r})$ induces a polarisation field $\mathbf{P}(\mathbf{r})$ which depends linearly on the field, provided the field is not too large. For such an insulating material, which we shall assume to be uniform and isotropic, we can write

▶ $$\mathbf{P}(\mathbf{r}) = \varepsilon_0\chi_\mathrm{e}\,\mathbf{E}(\mathbf{r}), \tag{10.11}$$

where χ_e is a number, characteristic of the material, called the *electric susceptibility*. The inclusion of ε_0 in (10.11) makes χ_e dimensionless.

As an example, the relation (10.3) for diffuse material suggests that for a dilute monatomic gas, $\chi_\mathrm{e} = 4\pi N\alpha$.

At high fields, the relation between $\mathbf{P}$ and $\mathbf{E}$ will become non-linear, and finally the insulator will break down through ionisation; the result may be spectacular, as in the case of lightning.

We shall call insulators for which (10.11) holds *dielectrics*. (The word was introduced by Faraday in 1838.) Other types of insulator exist. *Piezoelectric crystals* become electrically polarised when they are mechanically strained. A *ferroelectric crystal* has a spontaneous polarisation field $\mathbf{P}(T)$ at temperatures T below its characteristic 'ferroelectric Curie temperature' T_c; just above T_c the crystal is easy to polarise and hence has a very high susceptibility. This latter property of ferroelectrics is of some use technically in the construction of capacitors.

In the case of dielectrics

$$\blacktriangleright \qquad \mathbf{D} = \varepsilon_0 \mathbf{E} + \mathbf{P} = \varepsilon_0 \mathbf{E} + \varepsilon_0 \chi_e \mathbf{E} = \varepsilon_0 \varepsilon_r \mathbf{E}, \qquad (10.12)$$

where

$$\varepsilon_r = 1 + \chi_e \qquad (10.13)$$

is called the *relative permittivity* or *dielectric constant* (Table 10.1).

10.4 The surface of an insulator

At the surface of an insulator, $\mathbf{P}(\mathbf{r})$ falls from its bulk value $\mathbf{P}_i$ just inside the surface, to zero just outside the surface (Fig. 10.1). Integrating $\rho_b = -\boldsymbol{\nabla}\cdot\mathbf{P}$ through the volume ΔV between surfaces S_i just inside and S_0 just outside gives (with the help of the divergence theorem)

$$-\int_{\Delta V} \boldsymbol{\nabla}\cdot\mathbf{P}\,\mathrm{d}V = \int_{S_i} \mathbf{P}_i \cdot \mathrm{d}\mathbf{S} - \int_{S_0} \mathbf{P}\cdot \mathrm{d}\mathbf{S},$$

noting $\mathrm{d}\mathbf{S}_0 = \mathrm{d}\mathbf{S}$, $\mathrm{d}\mathbf{S}_i = -\mathrm{d}\mathbf{S}$.

Since $\mathbf{P}$ is zero over S_0, we see that the polarisation charge contained in the volume ΔV is equivalent to a 'surface charge' $\mathbf{P}_i \cdot \mathrm{d}\mathbf{S}$ associated with each element $\mathrm{d}\mathbf{S}$ of the surface of the insulator.

At the boundary between two insulators, the effect of the surface polarisation charges is to make the normal component of the $\mathbf{E}$ field discontinuous. However, the definition of $\mathbf{D}$ ensures that the value of $\mathbf{D}\cdot\mathrm{d}\mathbf{S}$ is the same on both sides, so that the normal component of $\mathbf{D}$ is continuous. The components of the $\mathbf{E}$ field parallel to the surface are also continuous. These results are proved in Appendix C.

In the interior of a uniform dielectric χ_e does not vary with position. Since $\boldsymbol{\nabla}\cdot\mathbf{E} = -\boldsymbol{\nabla}\cdot\mathbf{P}/\varepsilon_0$ by (10.7) and $\mathbf{P} = \varepsilon_0 \chi_e \mathbf{E}$, eliminating $\mathbf{E}$ gives $(1 + \chi_e^{-1})\boldsymbol{\nabla}\cdot\mathbf{P} = 0$, so that $\rho_b = -\boldsymbol{\nabla}\cdot\mathbf{P} = 0$. Thus in a uniform dielectric there is no bulk polarisation charge: all the polarisation charge lies at the surface.

Table 10.1. *The relative permittivity (dielectric constant) of some insulating materials*

Gases		$10^4(\varepsilon_r - 1)$
Air	(1 atm, 20 °C)	5.37
Argon	(1 atm, 20 °C)	5.16
Water vapour	(1 atm, 100 °C)	60
Water vapour	(1 mm Hg, 20 °C)	1.21
Liquids and solids		ε_r
Liquid argon	(82 K)	1.53
Water		80
Diamond	(20 °C)	5.5
Glass (Pyrex)		4–6
Polyethylene		2.3

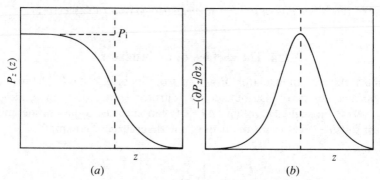

(a)   (b)

Fig. 10.1 (a) The behaviour of the component $P_z(z)$ of the polarisation field **P** at the surface of an insulator (taken as the xy plane), and (b) the corresponding contribution $(-\partial P_z/\partial z)$ to the bound charge.

10.5 Dielectrics in capacitors

It is instructive to understand the effect of a dielectric in a capacitor in terms of the polarisation charges. Consider a parallel plate capacitor in which the space between the plates is completely filled with a uniform dielectric medium, and suppose there to be surface charge densities $+\sigma$, $-\sigma$ on the inner surfaces of the metal plates. Since there is then an electric field between the plates, the dielectric is polarised in the direction of the field and surface charge densities $-P$, $+P$ are induced on the surfaces of the dielectric (Fig. 10.2). By Gauss's theorem the electric field across the plates is of magnitude

$$E = (\sigma - P)/\varepsilon_0.$$

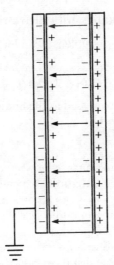

Fig. 10.2 Distribution of 'external' charge and bound polarisation charge in a parallel plate capacitor. The arrows indicate the electric field.

But for a dielectric, $P = \varepsilon_0 \chi_e E$, and hence

$$E(1 + \chi_e) = \sigma/\varepsilon_0, \text{ or } E = \sigma/\varepsilon_0 \varepsilon_r.$$

If the plates are distance d apart, the potential difference between them is

$$V = Ed = \sigma d/\varepsilon_0 \varepsilon_r,$$

and the capacitance per unit area is $\sigma/V = \varepsilon_0 \varepsilon_r/d$. Thus, filling the space between the plates with a dielectric increases the capacitance by a factor ε_r.

It is also instructive to see that the result may be obtained by making use of the **D** field we introduced above. When, as in this problem, there are external charges ρ_{ext}, **D** satisfies

$$\mathbf{\nabla} \cdot \mathbf{D} = \rho_{\text{ext}}.$$

In the uniform dielectric, $\mathbf{D} = \varepsilon_0 \varepsilon_r \mathbf{E}$, giving

$$\mathbf{\nabla} \cdot \mathbf{E} = \rho_{\text{ext}}/\varepsilon_0 \varepsilon_r.$$

Thus the electric field for a given charge on the plates is scaled down by a factor ε_r, compared with the corresponding vacuum problem, and the capacitance is thereby increased by a factor ε_r. Clearly this result holds not only for the parallel plate capacitor, but for *any* shape of capacitor in which the space between conductors is entirely filled with dielectric of uniform relative permittivity.

10.6 Electrostatics with dielectrics present

In time-independent systems involving conductors and insulators, the Maxwell equations are

$$\nabla \times \mathbf{E} = 0, \quad \nabla \cdot \mathbf{D} = \rho_{\text{ext}}. \tag{10.14}$$

The electric field can still be expressed as the gradient of a potential Φ, and with the constitutive relation $\mathbf{D} = \varepsilon_0 \varepsilon_r \mathbf{E}$,

$$\nabla \cdot (\mathbf{D}) = \nabla \cdot (\varepsilon_0 \varepsilon_r \mathbf{E}) = -\varepsilon_0 \varepsilon_r \nabla^2 \Phi = \rho_{\text{ext}}. \tag{10.15}$$

The boundary conditions at the surface between two dielectrics were given in §10.4. In analysing simple situations we can usually apply methods similar to those used in Chapter 7. As an example, we shall show how a uniform vacuum electric field is modified by the insertion of a dielectric sphere, of radius a and relative permeability ε_r (Fig. 10.3).

The polarisation charge induced on the sphere must give a dipole form of potential at sufficiently large distances. Let us assume (as in §7.4) that the potential outside the sphere is everywhere of the form

$$\Phi_{\text{out}}(r, \theta) = -E_0 r \cos \theta + \frac{\beta \cos \theta}{r^2},$$

where the term $-E_0 r \cos \theta = -E_0 z$ gives a constant field E_0 in the z-direction at large distances.

Inside the sphere we can anticipate that the potential will be of the form

$$\Phi_{\text{in}}(r, \theta) = \gamma r \cos \theta + (\text{constant}).$$

There can be no dipole term inside since this would imply an infinite field at the origin, which is unphysical.

At the surface of the sphere, $r = a$, the condition that the components of $\mathbf{E}$ parallel to the surface must be continuous is

$$\frac{1}{r} \frac{\partial \Phi_{\text{out}}}{\partial \theta} = \frac{1}{r} \frac{\partial \Phi_{\text{in}}}{\partial \theta}, \text{ giving } E_0 a - \frac{\beta}{a^2} = -\gamma a,$$

and equating the normal components of $\mathbf{D} = \varepsilon_0 \varepsilon_r \mathbf{E}$ we have

$$\frac{\partial \Phi_{\text{out}}}{\partial r} = \varepsilon_r \frac{\partial \Phi_{\text{in}}}{\partial r}, \text{ giving } -E_0 - \frac{2\beta}{a^3} = \varepsilon_r \gamma.$$

It is straightforward to solve these equations for β and γ, to obtain

$$\Phi_{\text{in}}(r, \theta) = -\left(\frac{3}{2 + \varepsilon_r}\right) E_0 r \cos \theta + (\text{constant}),$$

$$\Phi_{\text{out}}(r, \theta) = -E_0 r \cos \theta + \left(\frac{\varepsilon_r - 1}{\varepsilon_r + 2}\right) \frac{E_0 a^3 \cos \theta}{r^2}.$$

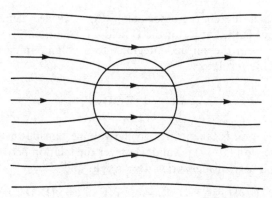

Fig. 10.3 Lines of the field **E** (or **D**) for a dielectric sphere with relative permittivity $\varepsilon_r = 4$ placed in an external uniform field.

Thus inside the sphere the **E** field is uniform and of magnitude $3E_0/(2 + \varepsilon_r)$. The corresponding polarisation is

$$\mathbf{P} = \varepsilon_0 \chi_e \, \mathbf{E} = \varepsilon_0(\varepsilon_r - 1) \, \mathbf{E} = \frac{3\varepsilon_0(\varepsilon_r - 1)}{(\varepsilon_r + 2)} \, \mathbf{E}_0,$$

so that the total dipole moment of the sphere, $\int \mathbf{P}\,dV$, is

$$\tfrac{4}{3}a^3\mathbf{P} = (4\pi\varepsilon_0)\, a^3 \left(\frac{\varepsilon_r - 1}{\varepsilon_r + 2} \right) \mathbf{E}_0.$$

This is exhibited in the dipole term in $\Phi_{\text{out}}(r, \theta)$. The uniqueness of our solution follows from extensions of the uniqueness theorem for the potential proved in Appendix B. (The constant appearing in Φ_{in} depends on the details of the surface structure, as in the case of a conductor.)

The simplicity of this example is perhaps a little misleading: an object of more complicated shape than a sphere can be expected to establish higher multipole fields which will be important close to the object.

10.7 Considerations on energy

The argument of §7.7 that the energy of a capacitor is increased by

$$\delta U = V\delta Q \tag{10.16}$$

when charge δQ is transferred from the conductor at potential $\Phi = 0$ to the conductor at potential $\Phi = V$, remains valid when insulating material is present. We can rewrite this result as

$$\delta U = \int \Phi \delta(\sigma\,dS) = \int_S \Phi\delta\mathbf{D}\cdot d\mathbf{S},$$

where the integral is taken over the surfaces of the conductors, and

$\sigma \, dS = \mathbf{D} \cdot d\mathbf{S}$ is the charge associated with a surface element $d\mathbf{S}$. This surface charge is the source of the $\mathbf{D}$ field (cf. equation (7.1)). As in §7.7, we can transform the surface integral into a volume integral, and since $\nabla \cdot \mathbf{D} = 0$ in the insulator

$$\delta U = \int_V \mathbf{E} \cdot \delta \mathbf{D} \, dV, \tag{10.17}$$

where the integral is taken through the volume outside the conductors.

The expression (10.17) is valid whatever the relation between $\mathbf{D}$ and $\mathbf{E}$. In the special case of dielectrics, $\mathbf{D} = \varepsilon_0 \varepsilon_r \mathbf{E}$, and

$$\mathbf{E} \cdot \delta \mathbf{D} = \varepsilon_0 \varepsilon_r \mathbf{E} \cdot \delta \mathbf{E} = \varepsilon_0 \varepsilon_r \delta(E^2/2) = \delta(\mathbf{E} \cdot \mathbf{D}/2).$$

We can then easily integrate (10.17) with respect to the field, to give

▶ $$U = \frac{1}{2} \int \mathbf{E} \cdot \mathbf{D} \, dV. \tag{10.18}$$

This is the energy relative to the state in which the capacitor is uncharged and the mean fields $\mathbf{E}$ and $\mathbf{D}$ are zero. We can regard this energy as residing in the dielectric, with energy density $(\mathbf{E} \cdot \mathbf{D})/2$. The expression (10.18) should be carefully distinguished from the general result we obtained in Chapter 2. The expression (2.26) is in terms of atomic fields, and includes the self energies of the constituent particles, and atomic potential energies which are part of their chemical binding energies. The expression $\mathbf{E} \cdot \mathbf{D}/2$ represents the additional energy density when the mean fields are not zero.

Problems

10.1   We noted in §3.1 that the atomic polarisability $4\pi\varepsilon_0 \alpha$ is of the order $4\pi\varepsilon_0 \times$ (atomic volume). Show that the electric susceptibility of an insulating solid made up of atomic units is of order unity, and that of a monatomic gas at STP is of order 10^{-4}. (See Table 10.1.)

10.2   According to statistical mechanics, for a dilute gas of polar molecules of electric dipole moment strength p, the probability distribution in angle θ of a moment with respect to an electric field E is

$$f(\theta) \, d\theta = \mathcal{N} \exp\left(pE\cos\theta/k_B T\right) \sin\theta \, d\theta,$$

when $\mathcal{N}$ is a normalisation constant.

Show that if N is the number density of the gas, the resulting mean polarisation, field is

$$P = Np[\coth\left(pE/k_B T\right) - (k_B T/pE)],$$

and at low fields this gives a contribution $Np^2/3\varepsilon_0 k_B T$ to the electric susceptibility.

10.3   The permanent electric dipole moment of a water molecule is 6.17×10^{-30} C m. Use the results of Problem 10.2 to estimate the electric

susceptibility of water vapour at temperature 20 °C, pressure 10 mm Hg
($N = 3.3 \times 10^{23}$ m^{-3}), and compare with the experimental value of
1.21×10^{-4}.

10.4   Find the relationship between the angles made with the normal when an
electric field passes through the surface between one medium, dielectric
constant ε_1 and another, dielectric constant ε_2.

10.5   A ferroelectric sphere of radius a has a uniform polarisation **P**. What is the
equivalent polarisation charge? Use the method of §10.6 to find the
potential, and hence the electric field, at points inside and outside the
sphere.

10.6   A 'cling-foil' film has thickness 25 μm, dielectric constant $\varepsilon_r = 3.25$ and
conductivity $\sigma = 10^{-16}$ Ω^{-1} m^{-1}. The film is sandwiched between two strips
of aluminium foil each 1 cm wide and 1 m long.

Show that the capacitance between the aluminium strips is 0.012 μF
and the resistance is 2.5×10^{13} Ω. With a potential difference between the
strips of 1 V, how many electrons pass through the cling-foil film per
second?

If the sandwich is rolled up with another cling-foil film in a spiral, does
the capacitance increase, decrease, or stay the same?

10.7   A charged and isolated parallel plate capacitor is half filled with an
insulating liquid of dielectric constant ε_r. If V_h and V_v are the potential
differences between the plates when they are horizontal and vertical
respectively, show that

$$\frac{V_h}{V_v} = \frac{(1+\varepsilon_r)^2}{4\varepsilon_r}.$$

10.8   A square parallel plate capacitor is of side L, and the sides are distance
$d \ll L$ apart. A square slab of dielectric also of side L, of thickness just less
than d, is inserted into the capacitor to a distance $b \gg d$, so that an area
bL lies inside the plates.

If the capacitor is isolated and carries charge Q, show that the electrical
energy is approximately

$$\text{energy} = \frac{Q^2}{2\varepsilon_0} \frac{d}{L^2 + (\varepsilon_r - 1)\,bL}$$

and deduce that there is a force acting to pull the dielectric into the
capacitor. What is the magnitude of this force?

Discuss the problem when the plates are maintained at a constant
potential difference V by a battery.

10.9   In a large capacitor, the electric field strength in the oil dielectric ($\varepsilon_r =$
2.24) is 9 MV m^{-1}. If there is a gas bubble in the oil, what is the field
strength in the bubble?

11

Magnetic properties of materials

Except for superconductors, all matter allows penetration of magnetic fields. In this chapter we outline the physical mechanisms which give rise to the magnetic properties of materials, and the phenomenological description of these properties in simple cases. Our discussion closely parallels that of electric polarisation in Chapter 10.

11.1 Magnetisation in materials

In §9.7 we saw that the orbital motion of the electrons in an atom, and the intrinsic magnetic moments of the electrons, may result in the atom having a magnetic moment. It is often the case that the atoms or molecules of a gas have electronic states which are filled in pairs in such a way that, in the absence of an external magnetic field, the net magnetic moment of an atom or molecule is zero. However, some atoms and molecules do possess unpaired electrons and have a permanent magnetic moment. In a gas of such atoms or molecules, the orientations of the magnetic moments will be randomised by thermal motion, but the moments will tend to align with an external magnetic field $\mathbf{B}$. This effect is called *paramagnetism*. A classical calculation for a dipole of magnitude μ gives its mean magnetic moment $\bar{\mu}$ to be

$$\bar{\mu} = \mu\left(\frac{\mu B}{3k_B T}\right), \tag{11.1}$$

at fields and temperatures such that $\mu|\mathbf{B}| \ll k_B T$. (Cf. Problem 10.2.)

Electrons also respond weakly in their orbital motion to an external magnetic field. In accord with 'Lenz's law' the induced atomic currents produce magnetic moments which are opposed to the external field (Problem 11.1). This *diamagnetism* is present in all atoms and molecules, and a gas of atoms or molecules which do not carry a permanent magnetic moment will usually be diamagnetic.

A similar discussion holds for paramagnetism and diamagnetism in liquids and solids. An additional phenomenon is that, in some crystals containing atoms with permanent magnetic moments, the electronic interactions are such that the moments may become spatially ordered even when no external magnetic field is present. *Ferromagnetism* is a special case of such ordering, and will be discussed in §11.6.

In all materials it is plausible that we can define a smooth averaged macroscopic *magnetisation field* $\mathbf{M}(\mathbf{r})$, such that the total magnetic moment of any macroscopic volume V is given by

▶
$$\text{magnetic moment} = \int_V \mathbf{M}(\mathbf{r})\,\mathrm{d}V. \tag{11.2}$$

For example, from (11.1), for a system of weakly coupled paramagnetic ions of number density $N(\mathbf{r})$ in an external magnetic field $\mathbf{B}(\mathbf{r})$,

$$\mathbf{M}(\mathbf{r}) = N(\mathbf{r})(\mu^2/3k_\mathrm{B}T)\,\mathbf{B}(\mathbf{r}). \tag{11.3}$$

11.2 The equivalent current distribution

To a macroscopic distribution of magnetisation (whether this is produced by atomic currents or intrinsic electron magnetic moments or both) there corresponds a macroscopic distribution of current. We use the mathematical identity

$$\int_V \mathbf{M}\,\mathrm{d}V = \frac{1}{2}\int_V \mathbf{r} \times (\nabla \times \mathbf{M})\,\mathrm{d}V, \tag{11.4}$$

where the surface S of the volume V completely encloses, and lies outside, the region of magnetic material.

This identity is easily proved by considering the components; for example, taking the x-component.

$$\int [y(\nabla \times \mathbf{M})_z - z(\nabla \times \mathbf{M})_y]\,\mathrm{d}V$$

$$= \int \left[y\frac{\partial M_y}{\partial x} - y\frac{\partial M_x}{\partial y} - z\frac{\partial M_x}{\partial z} + z\frac{\partial M_z}{\partial x} \right]\mathrm{d}V$$

$$= 2\int M_x\,\mathrm{d}V,$$

where in the last step we have integrated each term by parts, and the surface contributions vanish.

The magnetic moment of a current distribution $\mathbf{J}(\mathbf{r})$ is given by (9.12):

$$\int \mathbf{M}\,\mathrm{d}V = \frac{1}{2}\int \mathbf{r} \times \mathbf{J}(\mathbf{r})\,\mathrm{d}V. \tag{11.5}$$

Using (11.4) and (11.5) we can identify

▶ $$\mathbf{J}_{b}(\mathbf{r}) = \nabla \times \mathbf{M}(\mathbf{r}) \tag{11.6}$$

as the 'bound' current distribution corresponding to the magnetisation **M**.

The bound current must be included in Ampère's law to give the magnetic field in materials, along with any additional current $\mathbf{J}_{free}$ due to mobile charge carriers, so that for a system which is not varying with time (4.11) becomes

$$\nabla \times \mathbf{B} = \mu_0(\mathbf{J}_{free} + \mathbf{J}_{b}) = \mu_0(\mathbf{J}_{free} + \nabla \times \mathbf{M}). \tag{11.7}$$

When we were discussing electric polarisation we found it convenient to introduce the field **D**. Similarly here, we introduce the field **H(r)** defined by

▶ $$\mathbf{H} = (\mathbf{B}/\mu_0) - \mathbf{M}, \tag{11.8}$$

so that **H** satisfies

▶ $$\nabla \times \mathbf{H} = \mathbf{J}_{free}. \tag{11.9}$$

Integrating (11.9) over a surface S bounded by a curve Γ, and using Stokes's theorem, gives the integral form of this result:

$$\oint \mathbf{H} \cdot \mathbf{dl} = \int \mathbf{J}_{free} \cdot \mathbf{dS}. \tag{11.10}$$

11.3 Magnetic susceptibility and relative permeability

Except for those materials containing ordered paramagnetic ions, it is found experimentally that the magnetisation **M** of a material depends linearly on the field **B**, to a good approximation, so that **M**, **B**, and **H** are all proportional. If the material is uniform and isotropic, it is the convention to write

▶ $$\mathbf{M}(\mathbf{r}) = \chi_m \mathbf{H}(\mathbf{r}), \tag{11.11}$$

where the dimensionless constant χ_m is called the *magnetic susceptibility* of the material. (Since **H** is a 'derived' field, the reader may feel this convention is capricious, but it is rooted in the history of the subject.)

From (11.8), we then have

▶ $$\mathbf{B} = \mu_0(\mathbf{H} + \mathbf{M}) = \mu_0 \mu_r \mathbf{H}, \tag{11.12}$$

where

$$\mu_r = 1 + \chi_m \tag{11.13}$$

is called the *relative permeability*.

In solids which do not contain paramagnetic ions, diamagnetic

Table 11.1. *The magnetic susceptibilities at room temperature of some diamagnetic and paramagnetic materials*

Material	χ_m
Oxygen* (1 atm)	1.8×10^{-6}
Nitrogen (1 atm)	-0.63×10^{-8}
Water	-0.90×10^{-5}
Aluminium	2.2×10^{-5}
Copper	-0.96×10^{-5}
Germanium	-0.80×10^{-5}
Polyethylene	0.2×10^{-5}

* An oxygen molecule has a permanent magnetic dipole moment.

susceptibilities are small, $\sim -10^{-5}$. In metals there are also paramagnetic contributions to the susceptibility arising from the conduction electrons, which are of the same order of magnitude. (See Table 11.1.)

In systems containing disordered paramagnetic ions, it is still usually the case that $|\chi_m| \ll 1$, so that we may replace $\mathbf{H}$ by $\mathbf{B}/\mu_0$ in (11.12) to a good approximation. The classical result (11.3) then gives

$$\chi_m \approx \frac{C}{T} \qquad (11.14)$$

where $C = N\mu_0\mu^2/3k_B$. A $(1/T)$ dependence is indeed found experimentally in many such systems (Curie's law). For a solid with $N \approx 10^{28}$ m^{-3} and $\mu \approx \mu_B$ (the Bohr magneton: see §9.7), (11.14) gives $\chi_m \sim 10^{-3}$ at room temperature. This is of the order of the experimental values.

Thus we see that magnetic effects in materials which do not contain ordered paramagnetic ions are small, and for many purposes we may, when considering such materials, neglect them and set $\mu_r = 1$.

11.4 The surface of a magnetic material

At the surface of a magnetised material, the magnetisation falls from its bulk value $\mathbf{M}(\mathbf{r})$ just inside the surface, to zero outside the surface. As in §10.4, but now using (P.9), we may integrate the magnetisation current through the surface volume ΔV to obtain

$$\int_{\Delta V} \mathbf{\nabla} \times \mathbf{M} \, dV = -\int_{S_i} \mathbf{M} \times d\mathbf{S}_i = \int_S \mathbf{M} \times d\mathbf{S}.$$

Thus the magnetisation current in the surface region is equivalent to a 'surface current' of density $\mathbf{M} \times \hat{\mathbf{n}}$.

At the boundary between two materials, the normal component of the $\mathbf{B}$ field is continuous, since $\mathbf{\nabla} \cdot \mathbf{B} = 0$. The effect of the bound surface current is to make the components of the **B** field parallel to the surface discontinuous, but since $\nabla \times \mathbf{H} = \mathbf{J}_{\text{free}}$ the components of the **H** field parallel to the surface are continuous, provided there is no free surface current. (See Appendix C.)

11.5 Magnetic energy

Let us consider a long tightly wound solenoid of length l, cross-section S, with n turns per unit length, completely filled with uniform isotropic material. When a current I flows in the coil, the only significant **B** and **H** fields are in the material, and are parallel to the axis of the coil. The magnitude of **H** is $H = nI$ (cf. §9.5, but now $\oint \mathbf{H} \cdot \mathbf{dl} = \int \mathbf{J} \cdot \mathbf{dS}$). If I changes to $I + \delta I$, and B to $B + \delta B$, in time δt, the flux through the coil changes by $nlS\delta B$, and by Faraday's law a 'back e.m.f.' $-nlS\delta B/\delta t$ is induced. In time δt, the current source must do additional work $\delta U = I(nlS\delta B/\delta t)\delta t$ against this e.m.f. The volume of material is $V = Sl$, and $H = nI$, so that $\delta U = VH\delta B$. We can regard this energy as stored in the material; the change in magnetic energy per unit volume of material is $H\delta B$.

More generally it can be shown that the change in energy of a magnetic field when **B** changes to $\mathbf{B} + \delta \mathbf{B}$ is

$$\delta U = \int \mathbf{H} \cdot \delta \mathbf{B} \, dV. \tag{11.15}$$

We shall obtain this result, from a different point of view, in §12.2.

In the special case when $\mathbf{B} = \mu_r \mu_0 \mathbf{H}$, we can easily integrate (11.15) with respect to the field to give

$$\blacktriangleright \qquad U = \frac{1}{2} \int \mathbf{H} \cdot \mathbf{B} \, dV. \tag{11.16}$$

This is the magnetic energy relative to the magnetic energy of the material when the mean fields in it are zero.

11.6 Ferromagnetism

In a *ferromagnetic* crystal, over regions of the crystal called *domains* the magnetic moments of the ions are all essentially aligned in the same direction to give a spontaneous magnetisation $\mathbf{M}_0$. Both the magnitude and direction of $\mathbf{M}_0$ are constant within a domain, but the direction of $\mathbf{M}_0$ changes abruptly at domain boundaries. This domain structure makes the precise description of a sample difficult. The properties of a sample depend on its history, and experiments are rarely exactly reproducible.

If an external field is applied to a sample, domains having $\mathbf{M}_0$ favourably aligned with the field grow at the expense of other domains, and in stronger fields the $\mathbf{M}_0$ fields rotate, until at *saturation* the entire sample has its magnetisation aligned with the external field.

For engineering purposes, a magnetisation field $\overline{\mathbf{M}}$ can be defined on a very coarse scale δV, such that δV contains many domains. It is this average magnetisation which is of interest in device applications. The behaviour of $\overline{\mathbf{M}}(\mathbf{H})$ depends strongly on the ease, or difficulty, with which domain walls move. At saturation $|\overline{\mathbf{M}}| = |\mathbf{M}_0|$.

As an example, to obtain the observed low temperature saturation magnetisation of iron we must assign a moment $2.22\mu_B$ per atom. $|\mathbf{M}_0|$ is a function of temperature and vanishes above the *ferromagnetic Curie temperature* (1043 K for iron).

The magnetic field inside a domain depends on the shape of the domain and on the contribution of neighbouring domains. To obtain an idea of the order of magnitude of such fields we consider an isolated long cylindrical domain, of uniform magnetisation $\mathbf{M}_0$ parallel to the axis of the cylinder. Then $\nabla \times \mathbf{M}_0 = 0$ at interior points and the field is entirely due to the equivalent surface current $\mathbf{M}_0 \times \hat{\mathbf{n}}$. This vanishes on the ends of the cylinder, where $\mathbf{M}_0$ and $\hat{\mathbf{n}}$ are parallel, so that the magnetisation is equivalent to a bound surface current M_0 per unit length around the cylinder. We may thus use the arguments we developed for a solenoid in §9.5. Except near the ends, there is a field $\mathbf{B} = \mu_0 \mathbf{M}_0$ inside the domain (and hence, from (11.8), $\mathbf{H} \approx 0$). In iron this gives $|\mathbf{B}| = 2.18$ T, which is a substantial field.

As well as ferromagnets, there exist crystals with many other ordered arrangements of magnetic moment. These are discussed in texts on solid state physics. *Ferrites*, like ferromagnets, have a spontaneous magnetisation, but are insulators. They too have important applications in technology. For example, some types are used in transformer coils; since ferrites are insulators, eddy current losses are suppressed.

Problems

11.1   Consider an electron in an atom to be orbiting the nucleus on a circle of constant radius r with angular velocity ω. Show that if a magnetic field $\mathbf{B}(t)$ is gradually applied, normal to the plane of the orbit, the induced e.m.f. around the orbit changes ω by an amount

$$\Delta\omega = \frac{eB}{2m_e}.$$

Assume $\Delta\omega \ll \omega$.

Show that the change in the acceleration of the electron towards the nucleus is given by the magnetic force $e\mathbf{v} \times \mathbf{B}$ (supporting the assumption of constant r).

Show that the orbital magnetic moment of the electron changes by

$$\Delta m = -\left(\frac{e^2 r^2}{4m_e}\right)B.$$

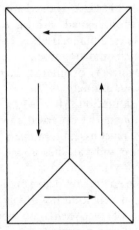

Fig. 11.1 The magnetisation **M** has the same magnitude in each domain, but different directions as shown.

11.2   Use the result of Problem 11.1 to estimate the magnetic susceptibility of a diamagnetic monatomic solid with Z electrons per atom and $N \sim 5 \times 10^{28}$ m^{-3}. Take $\overline{r^2} \sim (0.5 \text{ Å})^2$, for small Z.

11.3   A quantum mechanical calculation for an oxygen molecule gives $\mu = \sqrt{8}\mu_B$. Verify that the value for χ_m at room temperature which follows from equation (11.14) is in good agreement with experiment.

11.4   In the absence of external currents, (11.9) reduces to $\nabla \times \mathbf{H} = 0$. From (11.8), $\nabla \cdot \mathbf{H} = -\nabla \cdot \mathbf{M}$ (since $\nabla \cdot \mathbf{B} = 0$). As with the **E** field in electrostatics, **H** can therefore in this case be expressed as the gradient of a potential, $\mathbf{H} = -\nabla \Phi_m$ say, the source of which is $-\nabla \cdot \mathbf{M}$.

Use this observation to show that for a ferromagnetic sphere of uniform magnetisation **M**, radius a, inside the sphere

$$\mathbf{H} = -\tfrac{1}{3}\mathbf{M}, \quad \mathbf{B} = \tfrac{2}{3}\mu_0\mathbf{M},$$

and outside, the magnetic field is that of a dipole of strength $\tfrac{4}{3}\pi a^3 \mathbf{M}$. (Cf. Problem 10.5.)

11.5   A single rectangular crystal of iron has the domain structure and magnetisation **M** as indicated in Figure 11.1. Four of the domain boundaries are at 45° to **M**. By considering the sources of **H** (Problem 11.4) show that $\mathbf{H} = 0$ everywhere, $\mathbf{B} = 0$ outside the crystal, and inside $\mathbf{B} = \mu_0\mathbf{M}$. (The magnetic field energy is zero, a minimum, for this domain structure since $\mathbf{H} = 0$ everywhere.)

11.6   A ring of mean radius R, cross-section πr^2 where $r \ll R$, is made from material of high relative permeability μ_r. The ring is wound with N turns of wire which carry a current I. Estimate the magnitude of the B field inside the ring.

If a gap of length b is cut in the ring, where $b \ll r$, show that the magnetic field in the gap is approximately of magnitude

$$B_{\text{gap}} = \frac{N\mu_0\mu_r I}{(2\pi R - b) + b\mu_r}.$$

12

Time-dependent fields in insulators

In this chapter we consider time-dependent fields in insulators, and in particular the propagation of electromagnetic waves. We shall present a classical model to illustrate how a dielectric medium can respond resonantly to an oscillating electromagnetic field. The model will also illustrate how the medium can remove energy from, and hence attenuate, an electromagnetic wave. The dielectric 'constant' of Chapter 10 becomes a dielectric function $\varepsilon(\omega)$ of the frequency of oscillation ω of the field, with a real and an imaginary part.

12.1 The modified Maxwell equations

A time-dependent polarisation field $\mathbf{P}(\mathbf{r}, t)$ in an insulator corresponds to a time-dependent macroscopic charge density $\rho_b(\mathbf{r}, t) = -\nabla \cdot \mathbf{P}(\mathbf{r}, t)$, since the arguments of Chapter 10 remain valid if the time t is included as a parameter. Hence there must be a macroscopic current density $\mathbf{J}_b(\mathbf{r}, t)$. Charge and current are related by the continuity equation,

$$\frac{\partial \rho_b}{\partial t} + \nabla \cdot \mathbf{J}_b = 0,$$

giving

$$\frac{\partial}{\partial t}(-\nabla \cdot \mathbf{P}) + \nabla \cdot \mathbf{J}_b = 0.$$

Interchanging the order of differentiation,

$$\nabla \cdot \left(-\frac{\partial \mathbf{P}}{\partial t} + \mathbf{J}_b \right) = 0.$$

This implies that in general

$$\mathbf{J}_b = \frac{\partial \mathbf{P}}{\partial t} + \nabla \times \mathbf{M}, \tag{12.1}$$

99

where $\mathbf{M}(\mathbf{r}, t)$ is some vector field (cf. the introduction of the vector potential $\mathbf{A}$ in §9.1). One might guess that the contribution $\partial\mathbf{P}/\partial t$ comes from the dipole polarisation field, and a term of the form $\nabla \times \mathbf{M}$ comes from the magnetisation field $\mathbf{M}(\mathbf{r}, t)$ induced in the dielectric, as in Chapter 11. It is confirmed by atomic models that this assignment is essentially correct.

In insulators, and in the absence of 'external' charges and currents, the inhomogeneous Maxwell equations (5.7a) and (5.7b) therefore become

$$\nabla \cdot \mathbf{E} = \rho_{\mathrm{b}}/\varepsilon_0 = -\nabla \cdot \mathbf{P}/\varepsilon_0, \tag{12.2}$$

$$\nabla \times \mathbf{B} - \mu_0 \varepsilon_0 \frac{\partial \mathbf{E}}{\partial t} = \mu_0 \mathbf{J}_{\mathrm{b}} = \mu_0 \left(\frac{\partial \mathbf{P}}{\partial t} + \nabla \times \mathbf{M} \right).$$

These can be rewritten more concisely in terms of the auxiliary fields

$$\begin{aligned} \mathbf{D} &= \varepsilon_0 \mathbf{E} + \mathbf{P}, \\ \mathbf{H} &= (\mathbf{B}/\mu_0) - \mathbf{M}, \end{aligned} \tag{12.3}$$

and appear as homogeneous equations for these fields:

$$\blacktriangleright \qquad \nabla \cdot \mathbf{D} = 0, \quad \nabla \times \mathbf{H} - \frac{\partial \mathbf{D}}{\partial t} = 0. \tag{12.4}$$

The remaining Maxwell equations, (5.7c) and (5.7d), are immutable:

$$\blacktriangleright \qquad \nabla \cdot \mathbf{B} = 0, \quad \nabla \times \mathbf{E} + \frac{\partial \mathbf{B}}{\partial t} = 0. \tag{12.5}$$

The application of these equations depends on a knowledge of the constitutive relations. At sufficiently low frequencies we may suppose that the static relations are still valid. For a dielectric with $\mathbf{D} = \varepsilon_0 \varepsilon_r \mathbf{E}$, and $\mathbf{H} = (\mu_0 \mu_r)^{-1} \mathbf{B}$, (12.4) and (12.5) give

$$\nabla \times \mathbf{B} - \frac{\mu_r \varepsilon_r}{c^2} \frac{\partial \mathbf{E}}{\partial t} = 0,$$

$$\nabla \times \mathbf{E} + \frac{\partial \mathbf{B}}{\partial t} = 0, \tag{12.6}$$

where $c = (\mu_0 \varepsilon_0)^{-\frac{1}{2}}$ is the velocity of light in a vacuum.

Comparing these equations with the equations (6.1) for wave propagation in a vacuum, which were solved in Chapter 6, we see that similar solutions will be obtained, except that the velocity of propagation c is replaced by $c(\mu_r \varepsilon_r)^{-\frac{1}{2}}$.

12.2 Energy flow in insulators

Taking the scalar product of the second equation (12.4) with $\mathbf{E}$ gives

$$\mathbf{E} \cdot \nabla \times \mathbf{H} - \mathbf{E} \cdot \frac{\partial \mathbf{D}}{\partial t} = 0.$$

As in §6.5, we use the vector identity $\nabla \cdot (\mathbf{E} \times \mathbf{H}) = \mathbf{H} \cdot \nabla \times \mathbf{E} - \mathbf{E} \cdot \nabla \times \mathbf{H}$, and the Maxwell equation $\nabla \times \mathbf{E} + \partial \mathbf{B}/\partial t = 0$, to obtain

$$\mathbf{E} \cdot \frac{\partial \mathbf{D}}{\partial t} + \mathbf{H} \cdot \frac{\partial \mathbf{B}}{\partial t} + \nabla \cdot (\mathbf{E} \times \mathbf{H}) = 0.$$

Integrating this equation through a volume V enclosed by a surface S, and applying the divergence theorem to the last term, gives

$$\int_V \left(\mathbf{E} \cdot \frac{\partial \mathbf{D}}{\partial t} + \mathbf{H} \cdot \frac{\partial \mathbf{B}}{\partial t} \right) dV + \int_S (\mathbf{E} \times \mathbf{H}) \cdot d\mathbf{S} = 0. \qquad (12.7)$$

From §10.7, the term involving the integral of $\mathbf{E} \cdot \partial \mathbf{D}/\partial t$ gives the rate of change of the energy associated with the $\mathbf{E}$ field in the medium. From §11.5, the term involving the integral of $\mathbf{H} \cdot \partial \mathbf{B}/\partial t$ gives the rate of change of magnetic field energy. The surface integral gives the flow of energy across the surface S. Thus in an insulator, the Poynting vector becomes

▶ $$\mathbf{N} = \mathbf{E} \times \mathbf{H}, \qquad (12.8)$$

and we can interpret $\mathbf{N} \cdot d\mathbf{S}$ as the rate of flow of energy across an element of surface $d\mathbf{S}$. Equation (12.7) represents the conservation of energy.

For a dielectric medium with $\mathbf{D} = \varepsilon_0 \varepsilon_r \mathbf{E}$, and $\mathbf{H} = (\mu_0 \mu_r)^{-1} \mathbf{B}$,

$$\mathbf{E} \cdot \frac{\partial \mathbf{D}}{\partial t} = \tfrac{1}{2} \varepsilon_0 \varepsilon_r \frac{\partial}{\partial t} (\mathbf{E} \cdot \mathbf{E}) = \frac{1}{2} \frac{\partial}{\partial t} (\mathbf{E} \cdot \mathbf{D}),$$

and

$$\mathbf{H} \cdot \frac{\partial \mathbf{B}}{\partial t} = \tfrac{1}{2} (\mu_0 \mu_r)^{-1} \frac{\partial}{\partial t} (\mathbf{B} \cdot \mathbf{B}) = \frac{1}{2} \frac{\partial}{\partial t} (\mathbf{B} \cdot \mathbf{H}),$$

so that in this case the energy density of the electromagnetic field can be taken to be

$$\text{energy density} = \tfrac{1}{2} (\mathbf{E} \cdot \mathbf{D} + \mathbf{B} \cdot \mathbf{H}). \qquad (12.9)$$

This, we emphasise, gives the field energy relative to the energy when the mean fields $\mathbf{E}$ and $\mathbf{B}$ in the dielectric are zero.

12.3 The frequency dependent dielectric function $\varepsilon_r(\omega)$

We have seen in Chapter 10 and Chapter 11 that in most dielectric materials the interaction with an electric field is more significant than the interaction with a magnetic field, and we shall therefore here be particularly concerned with the polarisation $\mathbf{P}(t)$ of a uniform dielectric in a time-varying electric field.

Is it still possible to define a relative permittivity ε_r in a time-dependent situation? To address this question, we return to the simple classical model of an electron in an atom set up in §3.1, and consider the effect of a time-dependent external field of the form $\mathbf{E}(\mathbf{r}, t) = \mathbf{E}_0\, e^{i(\mathbf{k}\cdot\mathbf{r} - \omega t)}$. Since the equations we shall consider are all linear, we take a harmonic wave in complex form for mathematical convenience. We shall suppose that the wavelength of the field is much greater than the size of an atom, so that over an atom we can neglect the variation of $\mathbf{E}$ with position, and for an atom at the origin take $\mathbf{E} = \mathbf{E}_0\, e^{-i\omega t}$. This restriction of course excludes X-rays and γ-rays.

In the model of §3.1, when the electron cloud is displaced from the nucleus by $\mathbf{r}$ the restoring force on it is proportional to $\mathbf{r}$, and its equation of motion, neglecting the motion of the heavy nucleus, is

$$m_e \frac{d^2\mathbf{r}}{dt^2} = -m_e \omega_0^2 \mathbf{r} - m_e \gamma \frac{d\mathbf{r}}{dt} - e\mathbf{E}_0\, e^{-i\omega t}. \tag{12.10}$$

The restoring force $(-e^2/4\pi\varepsilon_0 a^3)\mathbf{r}$ of §3.1 has been written as $-m_e \omega_0^2 \mathbf{r}$, so that

$$\omega_0^2 = e^2/4\pi\varepsilon_0 a^3 m_e. \tag{12.11}$$

We have introduced a damping term $-m_e \gamma\, d\mathbf{r}/dt$ into the equation. In the absence of damping and with no external field, $d^2\mathbf{r}/dt^2 + \omega_0^2 \mathbf{r} = 0$, and the electron would oscillate indefinitely with frequency ω_0. The damping term leads to an exponential decay of these oscillations. Damping should always be expected, since an electron can always transfer energy to other electrons in the system or into radiation.

In the applied field the forced response of the electron is given by the solution of (12.10):

$$\mathbf{r} = -\frac{(e/m_e)}{(\omega_0^2 - \omega^2 - i\gamma\omega)}\mathbf{E}_0\, e^{-i\omega t}.$$

Hence the atom has an induced dipole moment

$$\mathbf{p}(t) = -e\mathbf{r}(t) = \frac{(e^2/m_e)}{(\omega_0^2 - \omega^2 - i\gamma\omega)}\mathbf{E}(t) = 4\pi\varepsilon_0\, \alpha(\omega)\, \mathbf{E},$$

where the atomic polarisability,

$$4\pi\varepsilon_0\, \alpha(\omega) = \frac{(e^2/m_e)}{(\omega_0^2 - \omega^2 - i\gamma\omega)} = 4\pi\varepsilon_0\, \alpha(0)\frac{\omega_0^2}{(\omega_0^2 - \omega^2 - i\gamma\omega)},$$

is now frequency dependent and, due to the damping, has an imaginary part. Writing

$$4\pi\varepsilon_0\, \alpha(\omega) = \frac{e^2}{m_e}\left[\frac{(\omega_0^2 - \omega^2) + i\gamma\omega}{(\omega_0^2 - \omega^2)^2 + \gamma^2\omega^2}\right],$$

we see that the imaginary part is positive; this has important implications for its interpretation.

The polarisation field of a sufficiently dilute gas of atoms of number density N will be given by

$$\mathbf{P}(\mathbf{r}, t) = 4\pi\varepsilon_0 N\alpha(\omega)\,\mathbf{E}(\mathbf{r}, t)$$

(cf. equation (10.3)), so that then

$$\mathbf{D} = \varepsilon_0\mathbf{E} + \mathbf{P} = \varepsilon_0\varepsilon_r(\omega)\,\mathbf{E}, \qquad (12.12)$$

where

$$\varepsilon_r(\omega) = 1 + \frac{(Ne^2/m_e\varepsilon_0)}{\omega_0^2 - \omega^2 - i\gamma\omega}. \qquad (12.13)$$

The characteristic forms of the real and imaginary parts of $\varepsilon_r(\omega) - 1$ in the neighbourhood of the resonant frequency ω_0 are shown in Fig. 12.1.

More generally we can expect that if an atom of the gas contains f_j electrons with natural frequencies ω_j and damping constants γ_j then

▶
$$\varepsilon_r(\omega) = 1 + \frac{Ne^2}{m_e\varepsilon_0}\sum_j \frac{f_j}{\omega_j^2 - \omega^2 - i\gamma_j\omega}. \qquad (12.14)$$

We have obtained this expression for the dielectric function of a dilute gas in a simple classical model. An expression of similar form is obtained in a quantum mechanical calculation, where however the ω_j correspond to transition frequencies to excited states of the atomic electrons, and the f_j become 'oscillator strengths'.

12.4 The dielectric function: molecular gases

The electronic transition frequencies ω_j in atoms and molecules which appear in the expression (12.14) correspond usually to energies $\hbar\omega_j$ of the order of electron volts, and hence give resonances lying in the visible or ultraviolet regions of the electromagnetic spectrum (Fig. 6.2). In molecules, the vibrational modes of oscillation of the ions may also interact with an oscillating electric field. Since the ions are of much greater mass than an electron, whilst the restoring forces acting are not dissimilar from those on an electron, the resonant frequencies are lower by a factor of

$$\sim \sqrt{\frac{\text{mass of electron}}{\text{mass of ion}}} \sim 10^{-2} \text{ or smaller (cf. (12.11)),}$$

and lie in the infrared. The dielectric function $\varepsilon_r(\omega)$ for a gas of molecules may have a complicated structure in the infrared as well as electronic resonances at higher frequencies. At lower frequencies (far infrared and microwave) rotational modes may be excited. The picture is further complicated by coupling between electronic, vibrational and rotational modes.

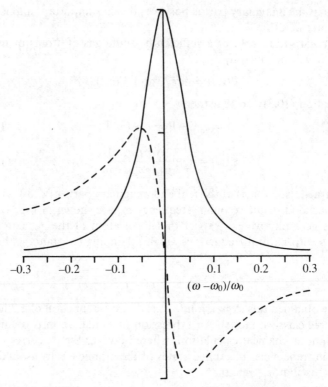

Fig. 12.1 The form of the real part (dotted line) and imaginary part (full line) of $\varepsilon_r(\omega) - 1$ in the neighbourhood of a resonance as given by equation (12.13), for $\gamma/\omega_0 = 10^{-1}$.

12.5 The imaginary part of $\varepsilon_r(\omega)$: attenuation

To understand the physical significance of a complex dielectric function, consider Maxwell's equations for fields varying with time as $e^{-i\omega t}$. Setting $\mu_r = 1$ for simplicity (and to a good approximation in most materials) we have, rewriting (12.4) and (12.5) appropriately,

$$\mathbf{\nabla}\cdot\mathbf{E} = 0 \quad \text{(a)}, \quad \mathbf{\nabla}\times\mathbf{B} - \frac{\varepsilon_r(\omega)}{c^2}(-i\omega\mathbf{E}) = 0 \quad \text{(b)},$$

$$\mathbf{\nabla}\cdot\mathbf{B} = 0 \quad \text{(c)}, \quad \mathbf{\nabla}\times\mathbf{E} + (-i\omega\mathbf{B}) = 0 \quad \text{(d)}. \tag{12.15}$$

We now seek travelling wave solutions of the form

$$\mathbf{E} = \mathbf{E}_0\, e^{i(\mathbf{k}\cdot\mathbf{r}-\omega t)},$$

and we find (cf. §6.4)

$$\mathbf{k}\cdot\mathbf{E}_0 = 0, \quad \mathbf{B} = (\mathbf{k}\times\mathbf{E})/\omega. \tag{12.16}$$

For simplicity we take $\mathbf{k}$ in the z-direction, $\mathbf{k} = (0, 0, k)$, and consider a linearly polarised wave with $\mathbf{E} = (E_x, 0, 0)$; then

$$E_x = E_0 \, e^{i(kz - \omega t)}, \tag{12.17}$$

and $\mathbf{B} = (0, B_y, 0)$, where

$$B_y = (kE_0/\omega) \, e^{i(kz - \omega t)}. \tag{12.18}$$

These fields satisfy (a), (c) and (d) above, and (b) is satisfied if $k^2 = \omega^2 \varepsilon_r / c^2$, that is

▶
$$k = \frac{\omega}{c} \sqrt{\varepsilon_r}. \tag{12.19}$$

(When $\varepsilon_r = 1$, this reduces to the vacuum relation $\omega = ck$.) Writing ε_r in terms of its magnitude and phase,

$$\varepsilon_r = |\varepsilon_r| \, e^{i\phi} = |\varepsilon_r| \cos\phi + i|\varepsilon_r| \sin\phi,$$

then

$$\sqrt{\varepsilon_r} = |\varepsilon_r|^{\frac{1}{2}} e^{i\phi/2} = |\varepsilon_r|^{\frac{1}{2}} \cos(\phi/2) + i|\varepsilon_r|^{\frac{1}{2}} \sin(\phi/2),$$

and we write this as

▶
$$\sqrt{\varepsilon_r} = n + i\kappa. \tag{12.20}$$

The imaginary part of ε_r is positive and hence the phase must lie in the range $0 \leqslant \phi \leqslant \pi$; thus both n and κ are positive numbers. n is called the *refractive index* of the medium.

Using (12.19) and (12.20), the electric and magnetic fields in the wave have

$$E_x = E_0 \, e^{i(\omega n/c)(z - ct/n)} e^{-\omega \kappa z/c},$$
$$B_y = (n + i\kappa)(E_x/c). \tag{12.21}$$

The actual fields are of course the real parts of these expressions. They represent a wave travelling in the z-direction with phase velocity $c/n(\omega)$, but the wave is also being attenuated. Both the phase velocity and the attenuation depend upon frequency. In optics this phenomenon is known as *dispersion*.

The energy flux in the wave is given by (12.8), which yields

$$|N| = |\varepsilon_r|^{\frac{1}{2}} (E_0^2/\mu_0 c) \cos[(\omega n/c)(z - ct/n)]$$
$$\times \cos[(\omega n/c)(z - ct/n) + \phi/2] \, e^{-2\omega \kappa z/c}.$$

Taking a time average over many oscillations,

$$\overline{|N|} = n(E_0^2/2\mu_0 c) e^{-2\omega \kappa z/c}, \tag{12.22}$$

(noting $n = |\varepsilon_r|^{\frac{1}{2}} \cos(\phi/2)$).

$K = 2\omega\kappa/c$ is called the *absorption coefficient*. Unless κ is zero, energy is lost from the wave as it propagates through the material.

12.6 The dielectric function: condensed matter

In condensed matter modes of oscillation which, in molecules, are distinct, merge and are spread into continuous bands. Much of the fine scale resonant structure characteristic of gases is lost. Expressions like (12.14), which are sums over individual frequencies, are replaced by integrals over continuous bands of frequencies. The complex dielectric function can still be defined and the refractive index and energy absorption coefficient can be measured experimentally.

In Fig. 12.2, the real and imaginary parts of the refractive index of hexagonal ice are displayed as functions of frequency, from microwave frequencies to the ultraviolet. (Much of the data comes from poly-crystalline samples, but the optical properties of a single ice crystal are approximately isotropic.)

In the infrared region a number of narrow resonances, corresponding to essentially intra-molecular vibrations, are evident, as well as broader resonances corresponding to crystal lattice vibrations. In the ultraviolet there is strong electronic absorption. Between the ionic resonances and the electronic resonances is a 'window' of low absorption, which is why ice is transparent to visible light. Such a window is apparent in many crystals and glasses. Another, and very striking, example is shown in Fig. 18.3, in the context of optical fibres. This window of low absorption in a very pure glass makes it possible to use optical fibres for the transmission of information over tens, or even hundreds, of kilometres.

12.7 Reflection from the plane boundary of a dielectric at normal incidence

Waves incident on the surface between two dielectrics are partially reflected and partially transmitted. Although the Maxwell equations in the media, together with the boundary conditions at the surface, determine the reflection and transmission of a wave in the general case, we shall here only consider a travelling wave at normal incidence from a vacuum region $z < 0$, onto the plane surface ($z = 0$) of a material with complex dielectric function $\varepsilon_r(\omega)$ and $\mu_r = 1$.

Using the results of §12.5, in the region $z < 0$ we expect the electric and magnetic fields to be a superposition of incident and reflected waves:

$$\mathbf{E} = \mathbf{E}_i \, e^{i(kz-\omega t)} + \mathbf{E}_r \, e^{i(-kz-\omega t)}$$
$$\mathbf{B} = (1/c)\,\hat{\mathbf{z}} \times \mathbf{E}_i \, e^{i(kz-\omega t)} - (1/c)\,\hat{\mathbf{z}} \times \mathbf{E}_r \, e^{i(-kz-\omega t)},$$

and in the dielectric material, $z > 0$,

$$\mathbf{E} = \mathbf{E}_t \, e^{i(k_t z-\omega t)}$$
$$\mathbf{B} = (\sqrt{\varepsilon_r}/c)\,\hat{\mathbf{z}} \times \mathbf{E}_t \, e^{i(k_t z-\omega t)}.$$

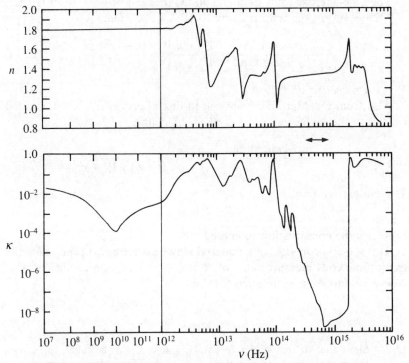

Fig. 12.2 The real and imaginary parts of the refractive index of ice at $-7\,°C$, from the microwave to the ultraviolet. The visible band is indicated by $\leftrightarrow$. Note how it overlaps the transparency window between ionic and electronic absorption. (Data from Warren, S. G. (1984), *Applied Optics* **23**, 1206).

The labels i, r, t refer to the incident, reflected, and transmitted waves. $\mathbf{E}_r$ and $\mathbf{E}_t$ are determined by the matching conditions at the interface. These conditions were stated for time-independent situations in §10.4 and §11.4. They hold also in the time-dependent case (Appendix C). The normal components of $\mathbf{D}$ and $\mathbf{B}$, and the tangential components of $\mathbf{E}$ and $\mathbf{H}$, are continuous. The first pair of conditions are trivially satisfied, since the normal components of the $\mathbf{E}$ and $\mathbf{B}$ fields are zero in this example. With $\mu_r = 1$ in the dielectric material for simplicity, the second pair give

$$\mathbf{E}_i + \mathbf{E}_r = \mathbf{E}_t,$$

$$\mathbf{E}_i - \mathbf{E}_r = \sqrt{\varepsilon_r}\,\mathbf{E}_t.$$

Hence for a given incident amplitude $\mathbf{E}_i$,

$$\mathbf{E}_r = \frac{1 - \sqrt{\varepsilon_r}}{1 + \sqrt{\varepsilon_r}}\,\mathbf{E}_i, \quad \mathbf{E}_t = \frac{2}{1 + \sqrt{\varepsilon_r}}\,\mathbf{E}_i. \tag{12.23}$$

The time-averaged flux of energy in the incident wave is, from (12.22), $|E_i|^2/2\mu_0 c$ and in the reflected wave is $|E_r|^2/2\mu_0 c$. Hence

$$R = \frac{\text{reflected flux}}{\text{incident flux}} = \left|\frac{1-\sqrt{\varepsilon_r}}{1+\sqrt{\varepsilon_r}}\right|^2 = \frac{(n-1)^2+\kappa^2}{(n+1)^2+\kappa^2}. \tag{12.24}$$

R is the *reflection coefficient*.

Also from (12.22), the time-averaged flux of energy in the transmitted wave into the surface at $z = 0$ is $n|E_t|^2/2\mu_0 c$, and

$$T = \frac{\text{transmitted flux}}{\text{incident flux}} = \frac{4n}{|1+\sqrt{\varepsilon_r}|^2} = \frac{4n}{(n+1)^2+\kappa^2}. \tag{12.25}$$

The evident relation,

$$R+T = 1,$$

expresses the conservation of energy.

The optical properties of a material may be determined experimentally by simultaneous measurements of R and the absorption coefficient $K = 2\omega\kappa/c$, as functions of frequency.

Problems

12.1 Show that at frequencies for which ε_r and μ_r can be taken to be real, the energy density U in a plane wave is equally divided between electric and magnetic contributions, and the Poynting vector is of magnitude $Uc/\sqrt{(\varepsilon_r \mu_r)}$.

12.2 Show that taking $a = 1$ Å in (12.11) yields a frequency in the ultraviolet.

12.3 From Fig. 12.2, the minimum value of κ for ice is $\kappa = 1.53 \times 10^{-9}$, at a frequency of 6.5×10^{14} Hz (blue). What is the corresponding attenuation length (i.e., inverse of the absorption coefficient)?
      What is the attenuation length in ice of microwave radiation at 2450 MHz (a typical frequency used in a microwave oven).

12.4 Light falls at normal incidence in a dielectric from a vacuum. Show that the reflected wave is retarded in phase by an angle $\tan^{-1}[2\kappa/(n^2+\kappa^2-1)]$.

12.5 Show that the reflection coefficient from a dielectric to a vacuum is the same as the reflection coefficient from a vacuum to the dielectric. (Take $\mu_r = 1$ in the dielectric).

12.6 Show that if $Re^{-KL} \ll 1$, a sample of thickness L transmits a fraction $(1-R)^2 e^{-KL}$ of the intensity of radiation falling on it at normal incidence.

12.7 *Snell's law.* The plane boundary $z = 0$ separates a medium ($z < 0$) with real dielectric constant ε_1 (and $\mu_r = 1$) from a medium ($z > 0$) with real dielectric constant ε_2 (and $\mu_r = 1$). For $z < 0$ there are solutions of Maxwell's equations which represent an incident and a reflected wave. If the electric field is of the form

$$\mathbf{E} = (0,1,0)(E_1 e^{ik_1 z} + E_r e^{-ik_1 z})e^{ik_2 x - i\omega t},$$

what is the magnetic field?

Show that the angle of incidence θ_i is given by $\tan\theta_i = k_2/k_1$, and if $\sqrt{(\varepsilon_1/\varepsilon_2)}\sin\theta_i < 1$ there is a transmitted wave with

$$\mathbf{E} = (0, 1, 0)\, E_t\, e^{ik_3 z}\, e^{ik_2 x - i\omega t}.$$

Writing $\tan\theta_t = k_2/k_3$, show that then

$$\sqrt{\varepsilon_1}\sin\theta_i = \sqrt{\varepsilon_2}\sin\theta_t. \quad \text{(Snell's law)}.$$

Find equations relating E_r and E_t to E_i.

12.8   *Total internal reflection.* Show that if in Problem 12.7 $\sqrt{(\varepsilon_1/\varepsilon_2)}\sin\theta_i > 1$, the electric wave for $z > 0$ is an *evanescent wave* of the form

$$\mathbf{E} = (0, 1, 0)\, E_t\, e^{-pz}\, e^{ik_2 x - i\omega t},$$

and all the energy of the incident wave is reflected.

12.9   If in Problem 12.7 the magnetic field for $z < 0$ is of the form

$$\mathbf{B} = (0, 1, 0)(B_i\, e^{ik_1 z} + B_r\, e^{-ik_1 z})\, e^{ik_2 x - i\omega t},$$

what is the electric field?

The magnetic field of the transmitted wave is of the form

$$\mathbf{B} = (0, 1, 0)\, B_t\, e^{ik_3 z}\, e^{ik_2 x - i\omega t}.$$

What is the electric field?

Show that if $\tan\theta_i = \sqrt{(\varepsilon_2/\varepsilon_1)}$ there is no reflected wave. The angle θ_i satisfying this condition is called the *Brewster angle.*

13

Time-dependent fields in metals and plasmas

13.1 Maxwell's equations for conductors

For a metal or a semiconductor, where there are mobile electrons present, the discussion of electric polarisation in Chapter 10 and Chapter 12 needs little modification. Motion of electrons bound in the ions and motion of the ions themselves can result in a polarisation field $\mathbf{P}_b$ and hence an associated charge density $\rho_b = -\nabla \cdot \mathbf{P}_b$ and current density $\mathbf{J}_b = \partial \mathbf{P}_b / \partial t$. There will be a contribution $\nabla \times \mathbf{M}$ to the bound current from magnetic effects. In addition, the mobile 'free' conduction electrons can give rise to a macroscopic current $\mathbf{J}_{free}$. If there is a macroscopic charge separation between the free electrons and the positive ions, there will be a macroscopic charge density ρ_{free} where

$$\rho_{free} = \langle \rho_{electrons} \rangle + \langle \rho_{ions} \rangle,$$

and the brackets denote the averaging of the atomic charge distributions. All of these charges and currents must be included in the macroscopic Maxwell equations, which now read:

$$\nabla \cdot \mathbf{D} = \rho_{free}, \tag{13.1a}$$

$$\nabla \times \mathbf{H} - \frac{\partial \mathbf{D}}{\partial t} = \mathbf{J}_{free}, \tag{13.1b}$$

▶
$$\nabla \cdot \mathbf{B} = 0, \tag{13.1c}$$

$$\nabla \times \mathbf{E} + \frac{\partial \mathbf{B}}{\partial t} = 0, \tag{13.1d}$$

where $\mathbf{D} = \varepsilon_0 \mathbf{E} + \mathbf{P}_b$ and $\mathbf{H} = (1/\mu_0)\mathbf{B} - \mathbf{M}$. (13.1a) and (13.1b) imply that ρ_{free} and $\mathbf{J}_{free}$ are related by the continuity equation

$$\frac{\partial \rho_{free}}{\partial t} + \nabla \cdot \mathbf{J}_{free} = 0. \tag{13.1e}$$

110

Similar equations hold in the case of a conducting liquid in which the mobile charge carriers are ions (an 'electrolyte'), and in the case of a plasma.

13.2 Conductors in oscillating fields

We now consider the response of a conductor to an oscillating electric field $\dot{\mathbf{E}}(\mathbf{r}, t) = \mathbf{E}_0 \, \mathrm{e}^{\mathrm{i}(\mathbf{k} \cdot \mathbf{r} - \omega t)}$. At sufficiently low frequencies, we can expect the steady state relation $\mathbf{J}_{\text{free}} = \sigma \mathbf{E}$ still to hold, at least approximately, but we shall see that, as the frequency ω increases, σ becomes a function of frequency, just as the dielectric function ε_r of an insulator becomes a function of frequency.

We use again the Drude model for free electrons, introduced in §8.1. The equation for the mean velocity $\bar{\mathbf{v}}$ of electrons in the neighbourhood of the origin $\mathbf{r} = 0$ is

$$m_e \frac{\mathrm{d}}{\mathrm{d}t} \bar{\mathbf{v}} = -m_e \gamma \bar{\mathbf{v}} - e\mathbf{E}_0 \, \mathrm{e}^{-\mathrm{i}\omega t}. \tag{13.2}$$

The damping term $-m_e \gamma \bar{\mathbf{v}}$ arises from electron collisions; it is clear that the equation holds in this simple form only if the spatial variation of the electric field is negligible over the mean distance an electron travels between collisions. This equation has the solution

$$\bar{\mathbf{v}} = \frac{-e\mathbf{E}_0 \, \mathrm{e}^{-\mathrm{i}\omega t}}{m_e(\gamma - \mathrm{i}\omega)}.$$

The corresponding current density is

$$\mathbf{J}_{\text{free}} = N(-e) \bar{\mathbf{v}} = \frac{Ne^2 \mathbf{E}}{m_e(\gamma - \mathrm{i}\omega)}, \tag{13.3}$$

where N is the number density of conduction electrons. When $\omega = 0$ the field is time-independent and (13.3) reduces to the result for the Drude model of §8.1, provided that we identify γ with the inverse of the collision time:

$$\gamma = \frac{1}{\tau}. \tag{13.4}$$

Since any point in the conductor might have been chosen as origin, we can write (13.3) as

$$\mathbf{J}_{\text{free}}(\mathbf{r}, t) = \sigma(\omega) \, \mathbf{E}(\mathbf{r}, t),$$

where

$$\sigma(\omega) = \frac{\sigma_0}{1 - \mathrm{i}\omega\tau}, \tag{13.5}$$

and $\sigma_0 = Ne^2\tau/m_e$ is the Drude static conductivity (8.4). The Maxwell equation (13.1b) becomes

$$\nabla \times \mathbf{H} - \frac{\partial \mathbf{D}}{\partial t} = \sigma(\omega)\,\mathbf{E}. \tag{13.6}$$

Note that $\sigma(\omega)$ is complex. Since, at frequency ω, $\partial \mathbf{E}/\partial t = -\mathrm{i}\omega\mathbf{E}$, we can rewrite (13.6) in the equivalent form

$$\nabla \times \mathbf{H} - \frac{\partial}{\partial t}\left(\mathbf{D} + \mathrm{i}\frac{\sigma(\omega)}{\omega}\mathbf{E}\right) = 0. \tag{13.7}$$

In (13.7), the response of the free electrons appears as a contribution to the displacement current, and corresponds to a contribution $\mathrm{i}\sigma(\omega)/\varepsilon_0\,\omega$ to the dielectric function. In fact (13.7) can be obtained directly from (12.14) by adding a term for the free electrons with $\omega_0 = 0$.

Thus at a non-zero frequency ω we may choose to interpret the physical response of the electrons to the electric field as due to a complex conductivity, or to a complex dielectric function. This observation allows us to apply the results of Chapter 12 to conductors by defining a complex dielectric function which includes the response of the conduction electrons:

$$\varepsilon_r(\omega) = \varepsilon_b(\omega) + \mathrm{i}\frac{\sigma(\omega)}{\varepsilon_0\,\omega}, \tag{13.8}$$

where $\varepsilon_b(\omega)$ is the dielectric function of the bound electrons.

13.3 Low frequency fields: $\omega\tau \ll 1$

Equation (13.5) shows that a significant frequency dependence in $\sigma(\omega)$ becomes manifest when $\omega\tau \sim 1$. For a good conductor at room temperature, $\tau \sim 10^{-14}$ s, so that up to, say, microwave frequencies $\sigma(\omega) \approx \sigma_0$. The contribution ε_b of the bound electrons to ε_r in (13.8), is then in most applications negligible in comparison with the contribution $\mathrm{i}\sigma_0/\varepsilon_0\,\omega$ from the conduction electrons, since ε_b is of order unity and $\sigma_0/\varepsilon_0\,\omega \gg 1$. (See Problem 13.1.)

Thus at low frequencies ($\omega \ll 10^{14}$ s^{-1}), it is usually a good approximation to retain only the contribution $\mathrm{i}\sigma/\varepsilon_0\,\omega$, and in the equations of §12.5 and §12.7 take $\varepsilon_r = \mathrm{i}\sigma_0/\varepsilon_0\,\omega$; we then have

$$\sqrt{\varepsilon_r} = n + \mathrm{i}\kappa = \sqrt{(\mathrm{i}\sigma_0/\varepsilon_0\,\omega)},$$

and since $\sqrt{\mathrm{i}} = (1+\mathrm{i})/\sqrt{2}$,

$$n = \kappa = \sqrt{(\sigma_0/2\varepsilon_0\,\omega)}. \tag{13.9}$$

From equation (12.21), a wave propagating in the conductor in the z-direction will be of the form (again taking $\mu_r = 1$):

$$\mathbf{E} = \mathbf{E}_0 \, e^{-z/\delta} \, e^{i(z/\delta - \omega t)}, \tag{13.10}$$

$$\mathbf{B} = \frac{(1+i)}{\omega \delta} \hat{\mathbf{z}} \times \mathbf{E}, \tag{13.11}$$

where

▶ $$\delta = \frac{c}{n\omega} = \sqrt{\left(\frac{2\varepsilon_0 c^2}{\sigma_0 \omega}\right)} = \sqrt{\left(\frac{2}{\mu_0 \sigma_0 \omega}\right)}. \tag{13.12}$$

Thus the amplitude of the wave falls by a factor $(1/e)$ over a distance δ.

In writing down (13.2), we made the proviso that we could neglect the spatial variation of the field over the mean distance between electron collisions. Hence the 'mean free path' of the electrons must be less than δ for (13.2) to be valid. This condition is well satisfied for $\tau \sim 10^{-14}$ s, $\omega\tau \ll 1$, but may break down for the much longer collision times characteristic of pure metals at very low temperatures, and a more elaborate theory is then necessary.

The reflection coefficient at low frequencies for a plane wave normally incident on a plane metallic surface from a vacuum is given by equation (12.24), and since $n = \kappa$ here,

$$R = \frac{(n-1)^2 + n^2}{(n+1)^2 + n^2} = \frac{(1-1/n)^2 + 1}{(1+1/n)^2 + 1} \approx 1 - \frac{2}{n}. \tag{13.13}$$

The last step comes from expanding in powers of $(1/n)$. For most metals at low frequencies $(\omega < 10^{11}$ s$^{-1})$, $n \gg 1$. Thus at low frequencies the wave is almost entirely reflected. The transmission coefficient T which gives the proportion of the incident beam lost into heating the conductor is

$$T = \frac{2}{n}, \tag{13.14}$$

since $R + T = 1$. The depth of penetration of the wave into the metal is characterised by the length δ, which is called the *skin depth*.

Whereas for plane waves in a vacuum $c|\mathbf{B}| = |\mathbf{E}|$, in a conductor (13.10) and (13.11) give

$$\frac{c|\mathbf{B}|}{|\mathbf{E}|} = \frac{c\sqrt{2}}{\omega\delta} = \sqrt{2}n \gg 1, \tag{13.15}$$

and there is a phase difference of $\pi/4$ between $\mathbf{E}$ and $\mathbf{B}$.

13.4 Alternating currents in wires

An analysis similar to that above can be carried out in a cylindrical geometry, and applied to the flow of alternating current in a wire of circular cross-section. In §8.5 we showed that in the case of steady currents the energy of Joule heating could be regarded as coming from the outside fields through the surface of the wire. The same is true for alternating currents. The associated oscillating fields only penetrate distances $\sim \delta$ into the wire. As the frequency increases, the flow of current becomes increasingly confined to the surface region of the wire. Thus for copper at room temperature, $(\sigma_0 = 5.8 \times 10^7 \ \Omega^{-1} \ m^{-1})$, at a mains frequency of $\omega = 2\pi \times 50 \ s^{-1}$, $\delta = 9.3$ mm; at a microwave frequency of $\omega = 2\pi \times 10^{10} \ s^{-1}$, $\delta = 0.66 \ \mu m$. For $\delta \ll$ (radius of wire) the results of the planar analysis become applicable (Problem 13.3).

13.5 The high frequency limit: $\omega\tau \gg 1$

Electromagnetic waves in the ultraviolet part of the spectrum have $\omega \sim 10^{16} \ s^{-1}$, and at these frequencies in most metals $\omega\tau \gg 1$. The contribution of the free electrons to the dielectric function $\varepsilon_r(\omega)$ becomes

$$\frac{i\sigma(\omega)}{\varepsilon_0 \omega} = \frac{i\sigma_0}{\varepsilon_0 \omega(1 - i\omega\tau)} \approx -\frac{\sigma_0}{\varepsilon_0 \omega^2 \tau} = -\frac{\omega_p^2}{\omega^2}, \tag{13.16}$$

where

$$\omega_p^2 = \frac{\sigma_0}{\varepsilon_0 \tau} = \frac{Ne^2}{\varepsilon_0 m_e}. \tag{13.17}$$

Note that ω_p, the *plasma frequency*, is independent of τ and depends only on the number density N of the conduction electrons. In most metals, $\omega_p \sim 10^{16} \ s^{-1}$.

In some metals at ultraviolet frequencies, $\chi_b(\omega)$ is small and $\varepsilon_b(\omega) \approx 1$, so that, using (13.16), the dielectric function (13.8) is real:

$$\varepsilon_r(\omega) \approx 1 - \frac{\omega_p^2}{\omega^2}. \tag{13.18}$$

For $\omega < \omega_p$, $\sqrt{\varepsilon_r}$ is, in this approximation, pure imaginary; putting $n \approx 0$ in (12.24), we see that $R \approx 1$. Thus a wave incident on the surface of the metal will be almost entirely reflected.

For $\omega > \omega_p$, $\sqrt{\varepsilon_r(\omega)}$ becomes real, giving

$$n \approx \sqrt{(1 - \omega_p^2/\omega^2)}, \quad \kappa \approx 0,$$

so that above its plasma frequency the metal should transmit electromagnetic waves with little attenuation (cf. (12.21)). Such is indeed the case for many simple metals, which become relatively transparent in the ultraviolet. Fig. 13.1 illustrates this.

13.6 Diffuse plasmas

A diffuse plasma is a gas in which some (or perhaps most) of the atoms and molecules are ionised, to form a diffuse medium of free electrons and ions. An important example of such a plasma is the *ionosphere*, which

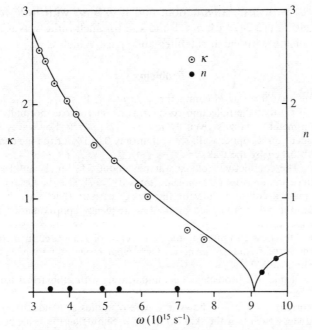

Fig. 13.1 The optical constants of sodium: the experimental points of κ and n are well fitted by the full curve, obtained from equation (13.18) with $\omega_p = 9.1 \times 10^{15} \text{ s}^{-1}$. (Data from Givens, M. P. (1958), *Solid State Physics* **6**, 313.)

extends between heights of about 60 km to 200–300 km above the earth's surface. The ionisation is caused principally by absorption of the sun's ultraviolet radiation, so that the density of free electrons and the electron collision time τ vary throughout the ionosphere. In particular the ionisation is dependent on altitude, and there are large diurnal variations.

To understand the gross properties of such a plasma the density of free electrons can be considered to be in the range 10^{10}–10^{12} m^{-3}, and the collision time $\tau \sim 10^{-4}$ s.

Taking $N = 10^{11} \text{ m}^{-3}$ in (13.17) gives a plasma frequency $\omega_p = 1.7 \times 10^7 \text{ s}^{-1}$, in the radio-frequency band and corresponding to a free space wavelength $\lambda \sim 100$ m. Around the plasma frequency, $\omega\tau \gg 1$, and the contribution to $\varepsilon_r(\omega)$ from the ions is negligibly small because of their large masses, and for the bound electrons $\varepsilon_b(\omega) \approx 1$, as in all dilute gases.

From the discussion in §13.5 above, electromagnetic waves with $\omega < \omega_p$, i.e., $\lambda > 100$ m, will be almost perfectly reflected, with little absorption. The presence of the ionosphere makes long distance radio communication possible: waves can be bounced between the ionosphere and the earth's surface to follow the curvature of the earth.

For $\omega > \omega_p$, and still with our crude model, the ionosphere transmits radio waves with little attenuation, and if $\omega \gg \omega_p$ with little reflection (using equation (12.24) with $n \approx 1$ and $\kappa \approx 0$). Such radio waves are used to communicate with earth satellites and space vehicles.

Problems

13.1   Show that for a good conductor, $\sigma_0/\varepsilon_0\omega \sim (10^{18} \text{ s}^{-1})/\omega$.

Show that the reflection coefficient for microwaves normally incident on a metal surface is given by $R = 1 - 2(2\varepsilon_0\omega/\sigma_0)^{\frac{1}{2}}$. For waves normally incident on a copper wall, show that only 0.015 % of the incident energy is absorbed by the wall.

13.2   (a)   The conductivity of sea water is about $5 \, \Omega^{-1} \text{ m}^{-1}$, and 1 kg of sea water contains about 0.5 mole of (ionised) NaCl. Use the Drude model to estimate a collision time for the charge carriers (Na^+ and Cl^-). Hence determine what is implied by the low frequency approximation ($\omega\tau \ll 1$) for sea water.

(b)   Obtain a formula for the skin depth of sea water in metres, as a function of frequency in hertz. Over what range of frequencies is your formula valid?

Consider the problems of communicating with submarines using radio waves.

13.3   A current $I = I_0 \cos\omega t$ flows in a wire of radius a, conductivity σ_0, at a frequency such that the skin depth $\delta \ll a$. Show that the Joule heating per unit length of the conductor is $\frac{1}{2}I_0^2/(2\pi a\sigma_0\delta)$.

13.4   Show that when $\omega\tau \gg 1$, the dispersion relation for waves in a plasma becomes

$$\omega^2 = \omega_p^2 + c^2k^2.$$

Find expressions for the group velocity v_g and phase velocity v_{ph}, and verify that

$$v_g v_{ph} = c^2.$$

13.5   When a space vehicle re-enters the Earth's upper atmosphere it ionises the atoms around it. If the electron density in this plasma is 10^{14} m^{-3}, what is the minimum frequency that should be used to maintain radio communication?

13.6   The (pulsed) radiation from a pulsar, at 300 MHz, is delayed 0.1 s with respect to radiation at 900 MHz. If the delay is caused by electrons in space, estimate the number of electrons in a 1 m² column along the line of sight. (Assume that the plasma frequency at all points along the line of sight is much less than 300 MHz.)

13.7   From the apparent plasma frequency for sodium (Fig. 13.1), estimate the number density of conduction electrons.

13.8   Show that for $\omega\tau \gg 1$, a more accurate expression for $\varepsilon_r(\omega)$ in a conductor is

$$\varepsilon_r(\omega) = 1 - \frac{\omega_p^2}{\omega^2} + \frac{i\omega_p^2}{\omega^3\tau},$$

and hence, above the plasma frequency, waves are attenuated with an absorption coefficient

$$\frac{2\omega\kappa}{c} = \frac{\omega_p}{\omega c\tau[(\omega/\omega_p)^2 - 1]^{\frac{1}{2}}}.$$

For sodium, $\tau = 3.2 \times 10^{-14}$ s (Problem 8.1). Estimate the attenuation length at twice the plasma frequency.

13.9 *Plasma oscillations.* In a plasma at high frequencies, equation (13.16) gives $\sigma(\omega) = i\varepsilon_0 \omega_p^2/\omega$. Show that, in such a plasma, charge density oscillations of the form $\rho_{\text{free}}(\mathbf{r}, t) = \rho_0(\mathbf{r}) e^{-i\omega t}$ are possible at the plasma frequency $\omega = \omega_p$.

14

Superconductors

In 1911 the Danish physicist Kamerlingh Onnes found that the electrical resistivity of mercury appeared to vanish completely below 4.2 K: a steady current flowed in a ring without need of a sustaining electric field. The discovery followed from his success in 1908 of liquifying helium, thereby making temperatures down to about 1 K accessible to experiment. This phenomenon of *superconductivity* was subsequently found in many other metals and alloys, but until recently the known *critical temperatures* T_c at which the transition to the superconducting state occurs had not exceeded 24 K; all superconducting technology depended on the availability of (expensive) liquid helium. In 1986–7 new classes of 'high-T_c' superconductors were discovered having critical temperatures which exceed the boiling point at atmospheric pressure of (cheap) liquid nitrogen, 77.4 K. These new superconductors are complex compounds, such as $YBa_2Cu_3O_{7-\delta}$; their properties and possible technological applications are being intensively studied.

14.1 The Meissner effect

Materials which become superconducting have the remarkable property of being 'perfectly diamagnetic' in their superconducting state. In a static magnetic field, up to a certain critical magnitude, magnetic flux is completely expelled from the inner regions of a large sample when the sample is cooled below its transition temperature (Fig. 14.1). An electric current is generated whose field exactly cancels the applied field in the interior of the sample. From the Maxwell equation $\nabla \times \mathbf{B} = \mu_0 \mathbf{J}$, the current flow is confined to a region close to the surface of the sample, since $\nabla \times \mathbf{B} = 0$ in its interior. Since the field is permanently excluded, the current must flow freely, without resistance. The effect was discovered by Meisner and Ochsenfeld in 1933 and can be regarded as the principle feature of superconductivity. It is not implied by infinite conductivity.

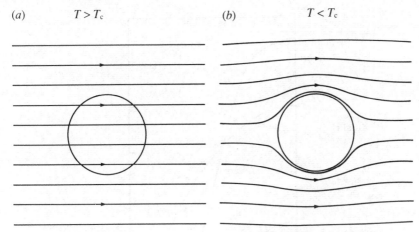

Fig. 14.1 Magnetic field lines in the neighbourhood of a super-conducting sphere placed in a uniform magnetic field, (a) above and (b) below its transition temperature. If the transition were simply to a state with $\sigma = \infty$, $\mathbf{E} = 0$, the magnetic field would remain unchanged, since then at all times $\partial \mathbf{B}/\partial t = -\nabla \times \mathbf{E} = 0$.

The perfect diamagnetism persists up to a critical field, denoted by B_c or B_{c1} depending on whether the superconductor is 'type I' or 'type II'; this distinction we shall explain later in the chapter. In a type I superconductor, the superconducting state is destroyed by a field $B > B_c(T)$, as is illustrated in Fig. 14.2. The magnetic properties of a type I superconductor are illustrated in Fig. 14.3.

14.2 The London equation

A very successful phenomenological theory of superconductivity was set up by Ginzberg and Landau in 1950, and the quantum mechanical model of Bardeen, Cooper and Schreiffer appeared in 1957. These theories are beyond the scope of this book, but we shall take from 'BCS theory' the basic notion that in a superconductor there is a collective, highly stable, quantum state, in which the superconducting electrons are bound in pairs: these are called 'Cooper pairs'. The mean velocity of the electrons in a Cooper pair is zero in a state in which no current flows.

A simple but useful description of superconductivity was introduced by the brothers F. and H. London in 1934. They suggested that in a superconductor a current density $\mathbf{J(r)}$ is spontaneously generated in response to any magnetic field $\mathbf{B(r)}$, related by the equation

$$\nabla \times (\mu_0 \Lambda^2 \mathbf{J}) = -\mathbf{B}, \qquad (14.1)$$

where Λ is a material and temperature dependent parameter.

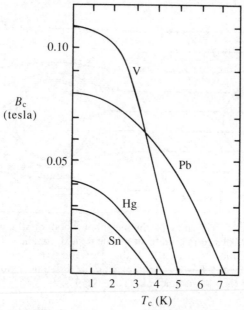

Fig. 14.2 The critical field B_c plotted against transition temperature T_c for some type I superconducting elements. The elements are superconducting in the regions below the curves. B_c is defined for the simple geometry of a long cylindrical sample with its axis parallel to the external field, so that end effects are negligible.

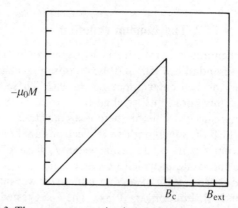

Fig. 14.3 The mean magnetisation M in a type I superconductor as a function of external field B_{ext}, for a long cylindrical sample parallel to the field. The continuity condition at the surface of the sample gives H (inside sample) $= B_{ext}/\mu_0$. Now $H = B/\mu_0 - M$. For $B < B_c$, $B = 0$ inside the sample, and hence $B_{ext} = -\mu_0 M$. In the normal state when $B > B_c$, $M \approx 0$.

This *London equation* supplements the Maxwell equations in a superconductor, in the same way as the equation $\mathbf{J} = \sigma\mathbf{E}$ relates current density and electric field and supplements the Maxwell equations in a normal conductor. We shall see that the Meissner effect follows from the London equation, and the equation also provides a basis for understanding the distinction between type I and type II superconductors.

To help elucidate (14.1), we consider moving a magnet close to a sample of superconductor. The magnetic field in the sample changes with time and, from Maxwell's equations, there will be an electric field. A Cooper pair will be accelerated by the mean Lorentz force acting on it. If $\mathbf{v}(\mathbf{r}, t)$ is the velocity of a Cooper pair centred on $\mathbf{r}$ at time t,

$$m_{\mathrm{C}}\frac{d\mathbf{v}}{dt} = -2e(\mathbf{E} + \mathbf{v} \times \mathbf{B}). \tag{14.2}$$

The pair carries charge $-2e$. The mass of the pair, m_{C}, will be an effective mass, of order of magnitude $2m_{\mathrm{e}}$. Since the resistance is zero there is no damping term. The acceleration of the pair is $(d\mathbf{v}/dt)$ and is the *total* time derivative.

Since the velocity field $\mathbf{v}(\mathbf{r}, t)$ depends on position,

$$\frac{d\mathbf{v}}{dt} = \frac{\partial\mathbf{v}}{\partial t} + \frac{\partial\mathbf{v}}{\partial x}\frac{dx}{dt} + \frac{\partial\mathbf{v}}{\partial y}\frac{dy}{dt} + \frac{\partial\mathbf{v}}{\partial z}\frac{dz}{dt} \tag{14.3}$$

$$= \frac{\partial\mathbf{v}}{\partial t} + (\mathbf{v} \cdot \mathbf{\nabla})\mathbf{v}.$$

Using equation (14.3) and the identity

$$\mathbf{\nabla}(\mathbf{v} \cdot \mathbf{v}) = 2(\mathbf{v} \cdot \mathbf{\nabla})\mathbf{v} + 2\mathbf{v} \times (\mathbf{\nabla} \times \mathbf{v}),$$

we can rewrite equation (14.2) as

$$\frac{\partial\mathbf{v}}{\partial t} + \tfrac{1}{2}\mathbf{\nabla}(v^2) + \frac{2e}{m_{\mathrm{C}}}\mathbf{E} = \mathbf{v} \times \left(\mathbf{\nabla} \times \mathbf{v} - \frac{2e}{m_{\mathrm{C}}}\mathbf{B}\right).$$

Taking the curl of this equation gives

$$\frac{\partial}{\partial t}\left(\mathbf{\nabla} \times \mathbf{v} - \frac{2e}{m_{\mathrm{C}}}\mathbf{B}\right) = \mathbf{\nabla} \times \left[\mathbf{v} \times \left(\mathbf{\nabla} \times \mathbf{v} - \frac{2e}{m_{\mathrm{C}}}\mathbf{B}\right)\right], \tag{14.4}$$

since $\mathbf{\nabla} \times \mathbf{\nabla}(v^2) \equiv 0$ and $\mathbf{\nabla} \times \mathbf{E} + \partial\mathbf{B}/\partial t = 0$.

It is clear that one solution of (14.4) is

$$\mathbf{\nabla} \times \mathbf{v} - \frac{2e}{m_{\mathrm{C}}}\mathbf{B} = 0. \tag{14.5}$$

If we start with the magnet far removed from the superconductor and with no supercurrent, so that $\mathbf{B} = 0$ and $\mathbf{v} = 0$, then initially

$[\nabla \times \mathbf{v} - (2e/m_C) \mathbf{B}] = 0$ everywhere in the sample, and equation (14.4) shows that this remains true at all subsequent times.

In terms of the number density of Cooper pairs N_C, the supercurrent density is

$$\mathbf{J}(\mathbf{r}, t) = -2eN_C \mathbf{v}(\mathbf{r}, t),$$

so that (14.5) may be written

$$\nabla \times \left(\frac{m_C}{4e^2 N_C}\right) \mathbf{J} + \mathbf{B} = 0. \tag{14.6}$$

If we identify

$$\blacktriangleright \qquad \Lambda = \left(\frac{m_C}{4\mu_0 e^2 N_C}\right)^{\frac{1}{2}}, \tag{14.7}$$

then (14.6) is identical with (14.1).

Equation (14.4) has other solutions. The London equation is that solution corresponding to a superconducting state.

14.3 Derivation of the Meissner effect

The Meissner effect follows from (14.1) with the help of the steady state Maxwell equation $\nabla \times \mathbf{B} = \mu_0 \mathbf{J}$. Eliminating $\mathbf{J}$,

$$\nabla \times (\Lambda^2 \nabla \times \mathbf{B}) = -\mathbf{B}. \tag{14.8}$$

In fields which are weak compared with B_c, we can neglect any spatial dependence of Λ (or, equivalently, N_C). Then, using the vector identity $\nabla \times (\nabla \times \mathbf{B}) = \nabla(\nabla \cdot \mathbf{B}) - \nabla^2 \mathbf{B}$ and noting $\nabla \cdot \mathbf{B} = 0$, (14.8) becomes

$$\nabla^2 \mathbf{B} = \Lambda^{-2} \mathbf{B}. \tag{14.9}$$

The Meissner effect is most easily displayed in a planar geometry. Consider a large slab of superconductor with a free surface at $z = 0$. Suppose the externally applied field is uniform and parallel to the surface, in the x-direction. Inside we can expect the field to be of the form $(B(z), 0, 0)$ so that equation (14.9) reduces to the ordinary differential equation

$$\frac{d^2 B}{dz^2} = \frac{1}{\Lambda^2} B.$$

The only solution of this equation which remains finite at all $z > 0$ is

$$B(z) = B_0 e^{-z/\Lambda}, \tag{14.10}$$

where B_0 is the field at the surface. Thus the field decays exponentially in the bulk. Λ is called the *London penetration depth*, and for a type I superconductor well below its critical temperature is typically 10^{-8}–10^{-7} m (10^2–10^3Å). This is small, but can be measured experimentally. The flux is negligible a few penetration depths from the surface.

The current density is similarly confined to the surface layer. Using $\nabla \times \mathbf{B} = \mu_0 \mathbf{J}$,

$$\mathbf{J} = (0, -(B_0/\mu_0 \Lambda)\,\mathrm{e}^{-z/\Lambda}, 0).$$

The 'surface current' density is of magnitude

$$\int_0^\infty J_y \, \mathrm{d}z = -B_0/\mu_0.$$

The London equation is an approximation to the results of more exact theories. In strong fields, the more sophisticated Ginzberg and Landau equations must be used. Nevertheless, even in strong fields the London equation gives a qualitative understanding of the phenomena.

14.4 The Abrikosov vortex

Another, and important, solution of equation (14.9) describes a cylindrical tube of magnetic flux surrounded by a vortex of supercurrent, all of which lies in the body of the superconductor. We shall see that penetration of these flux tubes can occur in the so-called 'type II' superconductors.

Consider solutions of the form

$$\mathbf{B} = (0, 0, B(\rho)),$$

where $\rho = (x^2 + y^2)^{\frac{1}{2}}$. (We are taking the flux tube to be along the z-axis of our coordinate system.) It is straightforward to show

$$\nabla^2 B = \frac{\mathrm{d}^2 B}{\mathrm{d}\rho^2} + \frac{1}{\rho}\frac{\mathrm{d}B}{\mathrm{d}\rho},$$

so that $B(\rho)$ satisfies

$$\frac{\mathrm{d}^2 B}{\mathrm{d}\rho^2} + \frac{1}{\rho}\frac{\mathrm{d}B}{\mathrm{d}\rho} = \frac{1}{\Lambda^2} B.$$

The solution of this equation which remains finite at large ρ is (constant) $\times K_0(\rho/\Lambda)$, where K_0 is a 'modified Bessel function' (Fig. 14.4). For $\rho \gg \Lambda$ the solution behaves like

$$\frac{1}{(\rho/\Lambda)^{\frac{1}{2}}}\,\mathrm{e}^{-\rho/\Lambda_{\mathrm{L}}},$$

which shows that the flux is confined in radius to a few penetration depths. The supercurrent,

$$\mathbf{J} = \mu_0^{-1}\nabla \times \mathbf{B} = \mu_0^{-1}\frac{\mathrm{d}B}{\mathrm{d}\rho}\left(\frac{y}{\rho}, -\frac{x}{\rho}, 0\right),$$

is clearly a vortex circling the flux tube and is similarly confined.

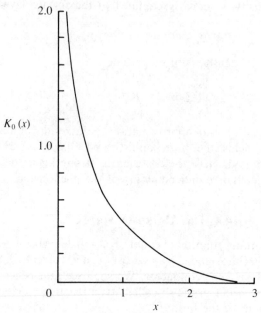

Fig. 14.4 The modified Bessel function $K_0(x)$. For $x \ll 1$, $K_0(x) \approx -\ln x$; for $x \gg 1$, $K_0(x) \approx (\pi/2x)^{\frac{1}{2}}e^{-x}$.

Although in this discussion we have used the London equation, our conclusions about the exponential decay are correct since the magnetic field is indeed small in the outer regions. It is only inside a small 'core' of the vortex where the field is large that our solution is flawed. The concept of a superconducting vortex was introduced by Abrikosov in 1958.

14.5 Flux quantisation and stability of flux tubes

Superconductivity is a macroscopic manifestation of a quantum mechanical state. A further remarkable manifestation is that the flux tubes which appear in type II superconductors are observed always to carry the same total magnetic flux:

▶ $$\Phi_0 = \int \mathbf{B} \cdot d\mathbf{S} = \frac{h}{2e},$$

where the integral is over a cross-section encompassing the flux, and h is Planck's constant; $(h/2e) \approx 2.068 \times 10^{-15}$ Wb.

A basic and very important property of flux tubes is that parallel tubes repel each other. This repulsion can be understood from a calculation of the energetics of flux tubes, which we shall not give. The force per unit length is exponentially small at separations $\gg \Lambda$ because of the confinement of flux and current, but at short distances stops two flux tubes coalescing into one (which would carry twice the quantum of flux). Anti-parallel flux tubes attract each other, and if they come sufficiently close will annihilate.

14.6 Type I and type II superconductors

It is the energetics of flux tubes which distinguishes type I from type II superconductors. Consider a long cylindrical superconducting sample (for which end effects will be negligible) enclosed in and placed parallel to the uniform magnetic field B produced by a solenoid which maintains a constant total flux across any cross-section.

We suppose that initially flux is excluded from the interior of the sample. If now a flux tube moves inside the sample, the magnetic energy per unit length of the field outside the sample decreases by

$$\frac{1}{2\mu_0}\int [B^2-(B-\delta B)^2]\,dS \approx \frac{B}{\mu_0}\int \delta B\,dS = \frac{B\Phi_0}{\mu_0}$$

where δB is the decrease in the magnetic field in the region outside the sample, and the integral is taken over the cross-section of the solenoid exterior to the sample. Since flux is conserved, $\int \delta B\,dS = \Phi_0$, and no extra work is done by the solenoid.

However, there is an energy, ε say, per unit length associated with a flux tube inside a superconductor, arising from the breaking of Cooper pairs and the magnetic and supercurrent field energy. If $B\Phi_0/\mu_0 < \varepsilon$ it is not energetically possible for a flux tube to move inside the sample. In type I superconductors $B_c\Phi_0/\mu_0 < \varepsilon$, so that flux tubes are never formed; the Meissner effect is always complete until at $B = B_c$ the superconducting state is destroyed.

In type II superconductors, there is a *lower critical field* B_{c1} for which $B_{c1}\Phi_0/\mu_0 = \varepsilon$; below B_{c1} flux is excluded. For $B > B_{c1}$, tubes will move in and will pack until their mutual repulsion makes more flux penetration energetically impossible. This can happen up to an *upper critical field* B_{c2}, when the whole sample is driven to the normal state. For $B_{c1} < B < B_{c2}$ the sample is said to be in a *mixed state*. It remains superconducting but the sample is permeated with tubes of flux (Fig. 14.5).

14.7 Hard superconductors

Whereas the critical field of a type I superconductor at liquid helium temperatures is typically $\sim 10^{-2}$ T (Fig. 14.2), the upper critical field B_{c2} of a type II superconductor can be very much higher. For example, in the case of the alloy Nb_3Sn (which has a critical temperature of 18.45 K), $B_{c2} \sim 22$ T. This greatly exceeds the values $B_{c2} \sim 0.2$ T for Nb, and $B_c \sim 0.03$ T for Sn!

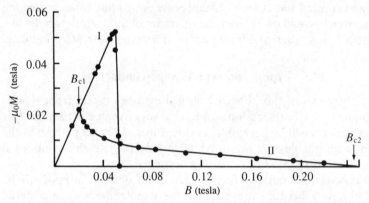

Fig. 14.5 The magnetisation curve of a type I superconductor (lead) compared with the magnetisation curve of a type II superconductor (lead – 8.23 wt % indium). (Data from Livingston, J. D. (1963), *Phys. Rev.* **129**, 1943.)

Nb$_3$Sn is one of the alloys used in the windings of superconducting magnets to produce high magnetic fields, thereby avoiding the problems of the dissipation of power which limit the performance of conventional magnets. However, such use is far from straightforward. When an increasingly high current is passed through a pure type II superconducting wire, flux tubes will enter the wire as soon as the external field reaches B_{c1}. These flux tubes are then driven by their interactions with each other and with the current until they either annihilate with an anti-parallel flux tube or pass out of the wire. In these processes they dissipate energy, and external power must be supplied to maintain the magnetic field. Thus the wire becomes resistive. However, the steady dissipation of energy through the motion of flux tubes is prevented if the flux tubes become 'pinned' at metallurgical defects, as apparently happens, so that a fixed structure of flux becomes established. 'Hard' type II superconductors can be manufactured which have a high density of suitable defects. These materials remain superconducting up to critical currents at which the Lorentz force on a flux tube exceeds the pinning force. Hard superconductors are used successfully in coils producing fields of up to ~ 18 T.

Problems

14.1   Estimate the maximum current that can be carried by a superconducting lead wire of radius 1 mm at 4 K. (See Fig. 14.2.)

14.2   A type I superconducting sphere is placed in a uniform external magnetic field $\mathbf{B}_{ext}$.
(a)   Show that if no flux enters it, the sphere has uniform magnetisation $\mathbf{M}$ where
$$-\mu_0 \mathbf{M} = \tfrac{3}{2}\mathbf{B}_{ext}.$$

(b)   Show that parts of the sphere become normal if $B_{ext} > \frac{2}{3}B_c$.
(Use the method of Problem 11.4. What is the boundary condition at the surface of the sphere?)

14.3   Consider the type II superconductor whose magnetisation curve is shown in Fig. 14.5. For fields between B_{c1} and B_{c2} the flux tubes in a long cylindrical sample placed parallel to the field, as viewed down their lengths, form an (equilateral) triangular lattice. Estimate the distance between adjacent flux tubes when the external field is 0.12 T.

   Assuming that only forces between adjacent tubes need be considered, estimate the repulsive force per unit length between flux tubes.

14.4   From conservation of total energy and conservation of total flux (no flux can pass through the superconductor), show that the magnetic field **B** inside a long superconducting solenoid exerts forces on the solenoid like a pressure $|\mathbf{B}|^2/2\mu_0$.

   Calculate this pressure in atmospheres for $B = 10$ T. (1 atm = 1.013×10^5 N m^{-2}.)

14.5   The resistivity of superconducting lead at 4.2 K has been measured to be less than 3.6×10^{-25} Ωm (Quin, D. J. and Ittner, W. B. (1962), *J. Appl. Phys.* **33**, 748). Estimate a lower limit for the decay time of a superconducting current in a lead ring, if the lead wire is of radius 0.1 mm, and the ring is of radius 10 mm.

15

Surface electricity

15.1 Contact potentials

In §7.1 we introduced the concept of the *work function* W, of a conductor in electrostatic equilibrium: W is the minimum work required to remove an electron from the interior of the conductor to a point just outside the conductor (where the macroscopic potential of the conductor is defined). The work function of a clean metallic surface is, typically, a few electron volts.

We argued in §7.7 that the work needed to remove an electron from one conductor at potential V_1 having work function W_1, to another at potential V_2 having work function W_2 is

$$\text{work} = (W_1 - W_2) + (-e)(V_2 - V_1). \tag{15.1}$$

If two initially uncharged conductors are brought together into close contact so that conduction electrons can flow between them, there will be in general a net transfer of charge until an equilibrium is reached when no work (positive or negative) is done on any electron making the transit. Since the equilibrium electrostatic field is conservative the work done in the transit does not depend on the path, and is still given by (15.1). Hence at equilibrium a potential difference

▶ $$V_{21} = V_2 - V_1 = (W_1 - W_2)/e \tag{15.2}$$

between the conductors will have been established by the flow of charge. Note that $V_{12} = -V_{21}$, and from (15.2) the contact potentials of any three conductors are related by

$$V_{12} + V_{23} + V_{31} = 0. \tag{15.3}$$

V_{21} is called the *contact potential* between the conductors. Contact potentials can be of the order of a volt (Table 15.1).

In the interior of each conductor the macroscopic potential will again

128

Table 15.1. *Selected values for the work function W of some metals*

Metal	Al	Au	Cs	Cu	Hg	Na	Pt	Th
W (eV)	4.28	5.1	2.14	4.65	4.99	2.75	5.65	3.4

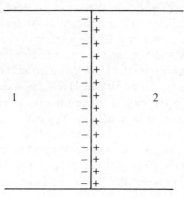

Fig. 15.1 Electronic charge transfer at the surface between two conductors, when $W_1 > W_2$.

be constant in equilibrium: the transferred charge will lie at the surface of the conductors, in a way determined by the macroscopic geometry of the system.

There will also be a change V_{21} in the difference between the internal macroscopic potentials of the two conductors in contact, and a related dipole moment $\varepsilon_0 V_{21}$ per unit area over the surface of contact (Fig. 15.1). This can be most easily seen if we model the averaged surface charge densities as two planes of charge densities σ and $-\sigma$ per unit area, distance d apart (cf. §7.6), and is true no matter how the charge at the interface is distributed (Problem 15.1). Taking $d = 1$ Å and $V_{21} = 1 V$ gives $\sigma = \varepsilon_0 V_{21}/d = 5.5 \times 10^{-3} e$ Å^{-2}: the fraction of conduction electrons transferred is very small and in metallic conductors these electrons can easily be provided from the immediate vicinity of the surface.

15.2 Metal–semiconductor junctions

The density of conduction electrons in most metals is $\sim 10^{-1}$ Å^{-3}. In contrast, typical semiconductors (for the most part specially manufactured materials) have much smaller carrier densities, in the range 10^{-10} Å^{-3} to 10^{-6} Å^{-3} at room temperature.

We shall consider a normal metal in contact with a semiconductor, and take the case in which the charge carriers in the semiconductor are electrons, and the work functions are such that electrons flow from the semiconductor into the metal.

When equilibrium is established, the transferred electronic charge will lie in the immediate vicinity of the contact, on the metallic side. However, because the semiconductor has such a low density of conduction electrons, there will be a *depletion layer* extending over a macroscopic distance, d say, in which to a first approximation there are no conduction electrons, and an electric field exists. Let us denote by $N_0 e$ the mean charge density of the positive ionic charge in the semiconductor, and let N be the number density of conduction electrons. In the interior of the semiconductor $N = N_0$ (since $\nabla \cdot \mathbf{E} = e(N_0 - N)/\varepsilon_0$, and in equilibrium $\mathbf{E} = 0$ in a region where conduction electrons are present). At the interface with the metal, all the conduction electrons in the depletion layer are swept out to provide the charge transferred to the metal, hence there is an electric field which in equilibrium will satisfy $\nabla \cdot \mathbf{E} = eN_0/\varepsilon_0$. Typically N_0 is so low that this insulating depletion layer has a thickness d of 10^3 Å to 10^5 Å. Thus it is indeed meaningful to define the macroscopic field $\mathbf{E}$ in the depletion layer.

Consider a plane interface with the z-axis normal to the interface (Fig. 15.2). In the depletion layer, $0 < z < d$, we have

$$\nabla \cdot \mathbf{E} = \frac{\partial E_z}{\partial z} = \frac{eN_0}{\varepsilon_0}$$

and $E_z = 0$ for $z > d$.

Hence for $0 < z < d$,

$$E_z = -\frac{\mathrm{d}V}{\mathrm{d}z} = eN_0(z-d)/\varepsilon_0,$$

since the field must be continuous at $z = d$ (where there is no layer of charge), and the potential $V(z)$ is

$$V(z) = -eN_0(z-d)^2/2\varepsilon_0.$$

In the metal $\mathbf{E} = 0$. The discontinuity across the surface at $z = 0$ arises from the layer of electronic charge at the surface. The contact potential V_{21} is related to the depletion layer thickness d by

$$V_{21} = -\int_0^d E_z\,\mathrm{d}z = eN_0\,d^2/2\varepsilon_0. \tag{15.4}$$

15.3 Rectification at a metal–semiconductor junction

Our analysis above has neglected thermal effects, which give a spread of energies to the conduction electrons both in the metal and in the semiconductor. In a more accurate treatment, we should allow for electrons in the metal which have sufficient thermal energy to pass into the

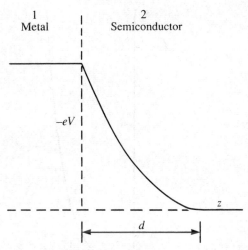

Fig. 15.2 The potential energy of an electron $(-e) V(z)$ in the neighbourhood of a metal–semiconductor interface due to the formation of the interface.

depletion layer, where the field will sweep them into the interior of the semiconductor. In equilbrium, this current will be balanced exactly by electrons in the semiconductor which have sufficient thermal energy to cross the potential barrier produced by the depletion layer. Statistical mechanics leads us to expect a Boltzmann factor $\exp(-eV_{21}/k_B T)$ in the expression for this latter current, which will be of the form $I_0 = A \exp(-eV_{21}/k_B T)$.

If now a small additional potential difference V is applied to the metal relative to the semiconductor, most of this potential will be sustained across the (nearly) insulating depletion layer. The Boltzmann factor in the expression for the semiconducting current will be modified to $\exp[-e(V_{21} - V)/kT]$, and the pre-factor will be little changed. The rate of detachment of electrons from the metal surface will also be little changed and they will still be swept into the semiconductor giving a current $-I_0$. Thus the total current will be

▶
$$I \approx I_0(e^{eV/k_B T} - 1). \tag{15.5}$$

I_0 is a characteristic of the particular junction.

Equation (15.5) gives a good representation of experimental data on the behaviour of current flow through the junction as a function of V (Fig. 15.3). The rectifying properties of the junction are evident.

The metal–semiconductor junction was an early example of a solid state electronic device, and was important in the early days of radio. It is today replaced by the '$p-n$ junction', in which similar rectification takes place between a 'p-type' semiconductor and an 'n-type' semiconductor.

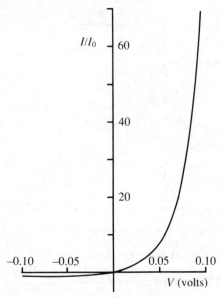

Fig. 15.3 I/I_0 from equation (15.5) at $T = 0\ °C$. (For fits of experimental data to expressions of this form see Arizumi, T. and Hirose, M. (1969), *Jap. J. of App. Phys.* **8**, 749.)

The resulting device, manufactured within a single crystal, is the key to much of modern solid state technology.

15.4 Galvanic cells

In thermal equilibrium, the potential difference $V_{21} = V_2 - V_1$ which is developed between two metals M_1 and M_2 in contact cannot be used to drive a current through a circuit: there is no source of energy. Indeed if a wire of metal M_3 is used to join M_1 and M_2, contact potentials V_{13} and V_{23} are set up such that the potential of each metal is constant; by (15.3), $V_{12} = V_{13} - V_{23}$. (If different parts of a system are maintained at different temperatures, currents can flow steadily, but we do not have space to embark on a discussion of thermoelectricity.)

In a *galvanic cell* ('battery') potential differences are similarly generated at the surfaces between different components of the cell, but there is a source of energy in the chemical reactions which take place. It is therefore possible for a current to be driven through an external circuit.

As an example, we consider the familiar lead–acid storage battery (invented by Planté in 1859). This consists of two electrodes dipping into a dilute solution of sulphuric acid (H_2SO_4). When the battery is fully charged, one electrode is of lead (Pb) and the other of lead coated with lead dioxide (PbO_2) over the area in contact with the electrolyte (Fig.

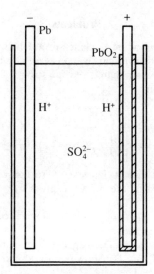

Fig. 15.4 A lead–acid battery (schematic).

15.4). In the lead and lead dioxide the charge carriers are electrons; in the electrolyte the carriers are H^+ and SO_4^{2-} ions. In static equilibrium dipole layers are formed at the electrode–electrolyte surfaces which make the lead electrode negative, and the coated electrode positive, with respect to the electrolyte. The e.m.f. of the cell is the potential difference between the two lead terminals. (There is a small contribution to the e.m.f. from the electronic Pb–PbO_2 contact potential.)

When the electrodes are joined by a wire, chemical reactions occur. At the lead–electrolyte interface,

$$Pb + SO_4^{2-} \to PbSO_4 + 2e^-,$$

and free electrons enter the metal, to be driven by the e.m.f. through the external circuit to the lead dioxide electrode, where they are captured by the reaction

$$PbO_2 + 4H^+ + SO_4^{2-} + 2e^- \to PbSO_4 + 2H_2O.$$

The overall chemical reaction that sustains the current on discharge is therefore

$$Pb + PbO_2 + 2H_2SO_4 \to 2PbSO_4 + 2H_2O.$$

The sulphuric acid concentration in the cell decreases, and lead sulphate ($PbSO_4$) is deposited on both electrode surfaces. These reactions are reversed by applying an external potential, greater than the e.m.f. of the cell and in the opposite sense, so that the battery may be recharged.

The above is a simple overview of what happens. In detail, the electrochemistry of a galvanic cell is a fascinating but complicated matter!

Problems

15.1   In the neighbourhood of a plane surface between two conductors, the averaged charge density can be considered to be of the form $\rho(z)$, with $\rho(z) \equiv 0$ for $|z|$ greater than a few ångströms. Show that the solution of Poisson's equation

$$\frac{\mathrm{d}^2 V}{\mathrm{d} z^2} = -\frac{\rho(z)}{\varepsilon_0}$$

is

$$V(z) = -\frac{1}{\varepsilon_0}\left[z \int_{-\infty}^{z} \rho(z')\,\mathrm{d}z' - \int_{-\infty}^{z} z' \rho(z')\,\mathrm{d}z' \right],$$

and provided that the net charge in the surface region is zero, the potential difference on crossing the surface is given by

$$\varepsilon_0 V_{21} = \int_{-\infty}^{\infty} z' \rho(z')\,\mathrm{d}z' = \text{dipole moment per unit area.}$$

15.2   The plates of a parallel plate capacitor are made of different metals with work functions W_1 and W_2. Show that the stored energy when the potential difference $V_2 - V_1 = V$ is

$$\text{energy} = C[V^2/2 + (W_1 - W_2)\,V/(-e)]$$

where $C = \varepsilon_0 A/d$ is the capacitance.
       Show this is a minimum when $V = (W_1 - W_2)/e$.

15.3   A parallel plate capacitor has plates of area 25 cm^2. One plate is made of platinum and the other of thorium. The plates are connected externally by a wire. Calculate the charge which flows round the wire if the distance between the plates is changed from 0.5 mm to 1 mm. (This effect was used by Lord Kelvin to infer contact potentials experimentally.)

15.4   The planar interface between two metals, platinum and thorium, is taken as the $x = 0$ plane, and the metals have a planar surface, taken as $y = 0$.
       Show that in the vacuum region $y > 0$, the potential that satisfies Laplace's equation and the boundary condition is

$$\Phi = \text{constant} + 0.72 \tan^{-1}(y/x)\ \text{V}.$$

Show that the electric field lines are semicircles centred on $x = y = 0$, and that the surface electron number density at a distance x Å along the platinum surface is given by

$$N = (4.0 \times 10^{-3}/x)\ \text{Å}^{-2}.$$

15.5   Consider a semiconductor having electron number density $N_0 = 10^{-8}$ Å^{-3} and a contact potential $V_0 = 1$ V above an adjacent normal metal. Show that, at the interface, the depletion layer in the semiconductor extends over $d \simeq 1000$ Å. Suppose that an additional potential bias $-V_b$ is applied to the metal so that d increases. Show that the charge per unit area on the metal surface becomes $\sigma = -[2\varepsilon_0 e N_0(V_0 + V_b)]^{\frac{1}{2}}$. The system can be used as a variable capacitor controlled by the bias voltage. Calculate the capacitance per unit area, $-\mathrm{d}\sigma/\mathrm{d}V_b$, for $V_b = 1$ V.

15.6   Suppose that on the discharge of a 2 V lead–acid cell 150 g of sulphuric acid (H_2SO_4) is consumed (typical for a 1 litre cell). Find the total charge (in ampere hours) passed between the terminals and estimate the mean electrical energy (in electron volts) liberated per H_2SO_4 molecule.

16

Radiation

In Chapter 6 we showed that Maxwell's equations have solutions which describe the propagation of electromagnetic waves in free space. In this chapter we develop the general theory relating electromagnetic fields to their sources; we defer examples of its application to Chapter 17.

16.1 The vector and scalar potentials

The Maxwell equation (5.7c), $\nabla \cdot \mathbf{B} = 0$, implies, as in §9.1, that we can introduce a *vector potential* $\mathbf{A}(\mathbf{r}, t)$ such that

▶
$$\mathbf{B}(\mathbf{r}, t) = \nabla \times \mathbf{A}(\mathbf{r}, t). \qquad (16.1)$$

The other homogeneous Maxwell equation (5.7d), $\nabla \times \mathbf{E} + \partial \mathbf{B}/\partial t = 0$ then becomes

$$\nabla \times \left(\mathbf{E} + \frac{\partial \mathbf{A}}{\partial t} \right) = 0.$$

We may therefore introduce a new scalar field $\Phi(\mathbf{r}, t)$ such that

$$\mathbf{E} + \frac{\partial \mathbf{A}}{\partial t} = -\nabla \Phi,$$

which gives

▶
$$\mathbf{E} = -\nabla \Phi - \frac{\partial \mathbf{A}}{\partial t}. \qquad (16.2)$$

$\Phi(\mathbf{r}, t)$ is the *scalar potential*. If the fields are static, Φ reduces to the familiar electrostatic potential.

Together, the potentials $\mathbf{A}(\mathbf{r}, t)$ and $\Phi(\mathbf{r}, t)$ specify the electromagnetic field. However, there is considerable freedom in the choice of these

135

potentials. Given A_1 and Φ_1, and any arbitrary function of space and time $\chi(\mathbf{r}, t)$, the potentials

$$A_2 = A_1 + \nabla\chi,$$

$$\Phi_2 = \Phi_1 - \frac{\partial\chi}{\partial t}, \tag{16.3}$$

generate exactly the same $\mathbf{E}$ and $\mathbf{B}$ fields (since $\nabla \times \nabla\chi \equiv 0$ and $\nabla(\partial\chi/\partial t)$ $= (\partial/\partial t)\,\nabla\chi$. This is a generalisation of §9.1 and is again called a *gauge transformation*.

In spite of this ambiguity, there is a considerable advantage in introducing these vector and scalar potentials, since the two homogeneous Maxwell equations are thereby satisfied identically. We may then substitute for $\mathbf{E}$ and $\mathbf{B}$ in the two inhomogeneous Maxwell equations (5.7a) and (5.7b), and use the freedom of choice allowed by a gauge transformation to simplify the resulting equations for $\mathbf{A}$ and Φ. In this book we shall impose the condition that

▶
$$\nabla \cdot \mathbf{A} + \frac{1}{c^2}\frac{\partial\Phi}{\partial t} = 0. \tag{16.4}$$

A choice of $\mathbf{A}$ and Φ which satisfies this condition is called a *Lorentz gauge* (Problem 16.1).

In a Lorentz gauge (5.7a) and (5.7b), with the help of the condition (16.4), give

$$\nabla \cdot \mathbf{E} = -\nabla^2\Phi - \nabla \cdot \left(\frac{\partial\mathbf{A}}{\partial t}\right) = -\nabla^2\Phi + \frac{1}{c^2}\frac{\partial^2\Phi}{\partial t^2} = \frac{\rho}{\varepsilon_0},$$

$$\nabla \times \mathbf{B} - \frac{1}{c^2}\frac{\partial\mathbf{E}}{\partial t} = -\nabla^2\mathbf{A} + \frac{1}{c^2}\frac{\partial^2\mathbf{A}}{\partial t^2} = \mu_0\,\mathbf{J}.$$

In obtaining the latter equation we have used the identity $\nabla \times (\nabla \times \mathbf{A}) = \nabla(\nabla \cdot \mathbf{A}) - \nabla^2\mathbf{A}$.

Thus in a Lorentz gauge $\mathbf{A}$ and Φ satisfy

▶
$$\nabla^2\Phi - \frac{1}{c^2}\frac{\partial^2\Phi}{\partial t^2} = -\frac{\rho}{\varepsilon_0},$$

$$\nabla^2\mathbf{A} - \frac{1}{c^2}\frac{\partial^2\mathbf{A}}{\partial t^2} = -\mu_0\,\mathbf{J}. \tag{16.5}$$

16.2 Radiation from moving charges: the general solution

The equations (16.5) for Φ and the three components of $\mathbf{A}$ are of similar form. Consider the equation for Φ:

$$\nabla^2\Phi - \frac{1}{c^2}\frac{\partial^2\Phi}{\partial t^2} = -\frac{\rho(\mathbf{r}, t)}{\varepsilon_0}. \tag{16.6}$$

Suppose first that the charge distribution has spherical symmetry about the origin, and is confined to the neighbourhood of the origin. We look for solutions $\Phi(r, t)$, which also have spherical symmetry. Away from the origin $\rho = 0$, and $\Phi(r, t)$ satisfies the wave equation

$$\nabla^2\Phi - \frac{1}{c^2}\frac{\partial^2\Phi}{\partial t^2} = 0. \tag{16.7}$$

Expressing the ∇^2 operator in polar coordinates, this becomes

$$\frac{1}{r}\frac{\partial^2}{\partial r^2}(r\Phi) - \frac{1}{c^2}\frac{\partial^2\Phi}{\partial t^2} = 0,$$

since we are assuming that Φ does not depend on the angular variables. We can write this as

$$\frac{\partial^2}{\partial r^2}(r\Phi) = \frac{1}{c^2}\frac{\partial^2}{\partial t^2}(r\Phi). \tag{16.8}$$

We now have an equation for $(r\Phi)$ of the same mathematical form as the 'one-dimensional wave equation'; this has the general solution

$$r\Phi = f\left(t - \frac{r}{c}\right) + g\left(t + \frac{r}{c}\right),$$

where f, g are two arbitrary functions. Hence

$$\Phi(r, t) = \frac{1}{r}f\left(t - \frac{r}{c}\right) + \frac{1}{r}g\left(t + \frac{r}{c}\right).$$

The first term clearly represents an outgoing spherical wave, and the second term an incoming spherical wave. In this chapter we shall be concerned with the production of radiation, rather than its absorption, so that we shall consider only

$$\Phi(r, t) = \frac{1}{r}f\left(t - \frac{r}{c}\right). \tag{16.9}$$

This solution satisfies the wave equation (16.7) everywhere except at the origin $r = 0$, where it is singular. Near the origin, expanding by Taylor's theorem,

$$\Phi(r, t) = \frac{1}{r}f(t) - \frac{1}{c}\frac{\partial f}{\partial t} + \frac{r}{2c^2}\frac{\partial^2 f}{\partial t^2} + \dots. \tag{16.10}$$

Now, in electrostatics, $Q/4\pi\varepsilon_0 r$ is the solution of Poisson's equation

$\nabla^2\Phi = -\rho/\varepsilon_0$ for a point charge Q at the origin, corresponding to $\rho(\mathbf{r}) = Q\delta(\mathbf{r})$. Hence we can write

$$\nabla^2(Q/4\pi\varepsilon_0 r) = -Q\delta(\mathbf{r})/\varepsilon_0,$$

which gives

$$\nabla^2(1/r) = -4\pi\delta(\mathbf{r}). \tag{16.11}$$

Thus when we use the expansion (16.10) close to the origin we obtain

$$\left(\nabla^2 - \frac{1}{c^2}\frac{\partial^2}{\partial t^2}\right)\left[\frac{f(t-r/c)}{r}\right] = -4\pi f(t)\,\delta(\mathbf{r}). \tag{16.12}$$

This identity is the key to finding a solution to the general problem! First we can move the point source of Φ from the origin to any point $\mathbf{r}'$, so that

$$\left(\nabla^2 - \frac{1}{c^2}\frac{\partial^2}{\partial t^2}\right)\left[\frac{f(t-|\mathbf{r}-\mathbf{r}'|/c)}{|\mathbf{r}-\mathbf{r}'|}\right] = -4\pi f(t)\,\delta(\mathbf{r}-\mathbf{r}'). \tag{16.13}$$

Then any charge distribution can be written as an integral over point sources:

$$\rho(\mathbf{r},t) = \int \rho(\mathbf{r}',t)\,\delta(\mathbf{r}-\mathbf{r}')\,\mathrm{d}V'. \tag{16.14}$$

If in (16.13) we set $4\pi f(t) = \rho(\mathbf{r}',t)/\varepsilon_0$, and integrate over $\mathbf{r}'$, we obtain the solution to (16.6) for the charge distribution $\rho(\mathbf{r},t)$:

$$\blacktriangleright \qquad \Phi(\mathbf{r},t) = \frac{1}{4\pi\varepsilon_0}\int\frac{\rho(\mathbf{r}',t-|\mathbf{r}-\mathbf{r}'|/c)}{|\mathbf{r}-\mathbf{r}'|}\,\mathrm{d}V'. \tag{16.15}$$

We can, similarly, construct a solution for each component of the equation for $\mathbf{A}$, and obtain

$$\blacktriangleright \qquad \mathbf{A}(\mathbf{r},t) = \frac{\mu_0}{4\pi}\int\frac{\mathbf{J}(\mathbf{r}',t-|\mathbf{r}-\mathbf{r}'|/c)}{|r-r'|}\,\mathrm{d}V'. \tag{16.16}$$

Our result (16.15) is a generalisation of the electrostatic potential (2.12), and (16.16) is a generalisation of the steady state vector potential (9.6). The Lorentz condition (16.4) is satisfied by our solutions, as may be checked by direct differentiation, together with the use of the continuity equation. (Care is needed!)

16.3 Electric dipole radiation

The expressions we have obtained above for $\mathbf{A}(\mathbf{r},t)$ and $\Phi(\mathbf{r},t)$ are exact for any charge and current distributions and for all $\mathbf{r}$ and t. However, we are often interested in the radiation out into free space from charge and current distributions confined to some finite region, where we may set the origin of coordinates. At large distances $\mathbf{r}$, it is then legitimate to take (cf. §3.1)

$$\frac{1}{|\mathbf{r}-\mathbf{r}'|} = \frac{1}{r} + \frac{\mathbf{r}\cdot\mathbf{r}'}{r^3} + \ldots \approx \frac{1}{r},$$

since even the second term in the expansion is of order r'/r^2, and is negligible at large r.

Hence at large r, we may replace (16.16) by

$$\mathbf{A}(\mathbf{r},t) = \frac{\mu_0}{4\pi r} \int \mathbf{J}(\mathbf{r}',t-|\mathbf{r}-\mathbf{r}'|/c)\,\mathrm{d}V'. \tag{16.17}$$

In the integrand, we have

$$t - \frac{|\mathbf{r}-\mathbf{r}'|}{c} = t - \frac{r}{c} + \frac{\mathbf{r}\cdot\mathbf{r}'}{cr} + \ldots. \tag{16.18}$$

Higher order terms in this expansion $\to 0$ as $r \to \infty$, but it may be necessary to retain the term $\mathbf{r}\cdot\mathbf{r}'/cr$, which is of order r'/c. This is of the magnitude of the time taken for light to cross the current distribution, and may be comparable with the time over which the current distribution changes appreciably, so that we should in general take

$$\mathbf{A}(\mathbf{r},t) = \frac{\mu_0}{4\pi r} \int \mathbf{J}\left(\mathbf{r}',t-\frac{r}{c}+\frac{\hat{\mathbf{r}}\cdot\mathbf{r}'}{c}\right)\mathrm{d}V'. \tag{16.19}$$

However, if on this time scale the current distribution is slowly varying, it is a good approximation to set

$$\mathbf{A}(\mathbf{r},t) = \frac{\mu_0}{4\pi r} \int \mathbf{J}\left(\mathbf{r}',t-\frac{r}{c}\right)\mathrm{d}V'. \tag{16.20}$$

The integral in (16.20) may be written in terms of the electric dipole moment $\mathbf{p}$ of the charge distribution. Consider the x-component of the dipole moment,

$$p_x(t) = \int x\rho(r,t)\,\mathrm{d}V.$$

Then

$$\frac{\mathrm{d}p_x}{\mathrm{d}t} = \int x\frac{\partial\rho}{\partial t}\,\mathrm{d}V$$

$$= -\int x\nabla\cdot\mathbf{J}\,\mathrm{d}V, \text{ using the continuity equation}$$

$$= \int J_x\,\mathrm{d}V. \tag{16.21}$$

In the final step, we have used the identity

$$\nabla \cdot (x\mathbf{J}) = J_x + x\nabla \cdot \mathbf{J},$$

and the divergence theorem: the surface terms vanish since the current distribution is zero at large distances.

▶ Thus
$$\mathbf{A}(\mathbf{r}, t) = \frac{\mu_0}{4\pi r} \dot{\mathbf{p}}\left(t - \frac{r}{c}\right). \tag{16.22}$$

The magnetic field at large r is

$$\mathbf{B} = \nabla \times \mathbf{A} = -\frac{\mu_0}{4\pi c} \frac{\hat{\mathbf{r}} \times \ddot{\mathbf{p}}(t - r/c)}{r} \tag{16.23}$$

(since the terms given by the ∇ operator are like, for example,

$$\frac{\partial}{\partial x} \dot{p}_y(t - r/c) = \ddot{p}_y \frac{\partial}{\partial x}\left(-\frac{r}{c}\right) = -\frac{x}{rc}\ddot{p}_y,$$

and we have neglected the term of higher order which comes from differentiating the $(1/r)$ factor in the expression (16.22) for $\mathbf{A}$).

We do not need to calculate $\Phi(\mathbf{r}, t)$ to find $\mathbf{E}$, since we can, more conveniently, use the Maxwell equation

$$\nabla \times \mathbf{B} - (1/c^2)\,\partial \mathbf{E}/\partial t = 0.$$

To leading order in $(1/r)$, we find

$$\mathbf{E} = -\frac{\mu_0}{4\pi r}[\ddot{\mathbf{p}} - \hat{\mathbf{r}}(\hat{\mathbf{r}} \cdot \ddot{\mathbf{p}})] \tag{16.24}$$
$$= c\mathbf{B} \times \hat{\mathbf{r}}.$$

This result is in fact evident without calculation: at large distances a spherical wavefront will appear locally to be plane, and the $\mathbf{E}$ and $\mathbf{B}$ fields will be perpendicular to the direction of propagation $\hat{\mathbf{r}}$, and to each other (cf. §6.2). The relationships (16.23) and (16.24) are illustrated in Fig. 16.1. In this figure, the z-axis has been taken in the direction of $\ddot{\mathbf{p}}$. Provided that $\ddot{\mathbf{p}}$ is not zero, the radiation is said to be *electric dipole*.

The power radiated per unit area in the direction $\mathbf{r}$ at time t is given by the Poynting vector:

$$\mathbf{N} = \mu_0^{-1}\mathbf{E} \times \mathbf{B} = (c/\mu_0)\,|\mathbf{B}|^2\hat{\mathbf{r}},$$

or

▶
$$\mathbf{N} = \frac{\mu_0}{(4\pi)^2 cr^2}\,|\ddot{\mathbf{p}}(t - r/c)|^2 \sin^2\theta\,\hat{\mathbf{r}}, \tag{16.25}$$

where the $\sin^2\theta$ comes from the vector product in (16.23). Energy flows

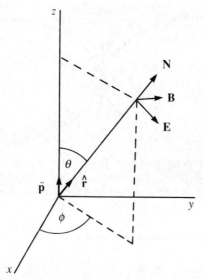

Fig. 16.1 Electric dipole radiation: the relative directions of $\ddot{\mathbf{p}}(t-r/c)$, **E, B** and **N**, at a distant point **r**.

away from the source with the velocity of light: the time delay r/c between source and point of observation is manifest, in that $\mathbf{N}(\mathbf{r}, t)$ depends on **p** evaluated at the earlier time $(t-r/c)$.

A *polar diagram* of the radiated power is shown in Fig. 16.2. The pattern has axial symmetry about Oz. The energy per unit time radiated in a solid angle $d\Omega$ is $|\mathbf{N}|\, r^2\, d\Omega$, and integrating over all angles we find energy is lost from the source at a rate

$$\frac{d\mathscr{E}}{dt} = \frac{\mu_0}{(4\pi)^2 c}|\ddot{\mathbf{p}}|^2 \iint \sin^2\theta \sin\theta\, d\theta\, d\phi$$

▶ (16.26)

$$= \frac{2}{3}\frac{1}{4\pi\varepsilon_0}\frac{1}{c^3}|\ddot{\mathbf{p}}|^2.$$

If $\ddot{\mathbf{p}}(t)$ vanishes, we must expand the integrand in (16.19) in powers of $\hat{\mathbf{r}}\cdot\mathbf{r}'/c$. Each term gives rise to a characteristic radiation pattern, and, provided that the source is small, the first non-vanishing term dominates. We shall consider only electric dipole radiation.

Problems

16.1   Suppose that in a certain gauge the potentials $\mathbf{A}_1$, Φ_1, satisfy

$$\nabla\cdot\mathbf{A}_1 + \frac{1}{c^2}\frac{\partial\Phi_1}{\partial t} = F(\mathbf{r}, t) \neq 0.$$

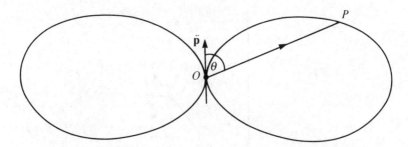

Fig. 16.2 The polar diagram for electric dipole radiation. The radiation has axial symmetry about the direction of $\ddot{\mathbf{p}}$. The power per unit solid angle flowing in a direction OP making an angle θ with $\ddot{\mathbf{p}}$, is proportional to the length OP.

Show that, with the gauge transformation (16.3),

$$\nabla \cdot \mathbf{A}_2 + \frac{1}{c^2}\frac{\partial \Phi_2}{\partial t} = 0$$

if χ is chosen to be

$$\chi(\mathbf{r}, t) = \frac{1}{4\pi}\int \frac{F(\mathbf{r}', t - |\mathbf{r}-\mathbf{r}'|/c)}{|\mathbf{r}-\mathbf{r}'|}\, dV'.$$

16.2   In §12.3 we introduced the model of an electron bound harmonically in an atom. Neglecting damping, $m_e\ddot{\mathbf{r}} = -m_e\omega_0^2\mathbf{r}$, and the electron can oscillate linearly with frequency ω_0; $\mathbf{r} = \mathbf{r}_0\cos\omega_0 t$.
   (a)   Show that the energy of oscillation is $E = m_e\omega_0^2 r_0^2/2$.
   (b)   Show that radiation is emitted at the same frequency as the oscillator frequency.
   (c)   Show that the dipole radiation formula should be a good approximation if $\lambda_0 \gg r_0$, where λ is the wavelength of the emitted radiation.
   (d)   Show then that, when averaged over many oscillations, the electron loses energy according to

$$dE/dt = -E/\tau,$$

   where the time constant τ is given by

$$\tau = \frac{3}{2}\left(\frac{4\pi\varepsilon_0}{e^2}\right)\left(\frac{m_e c^3}{\omega_0^2}\right).$$

16.3   An electric dipole of constant strength p_0 rotates about a perpendicular axis through its centre, at constant angular velocity ω_0. Derive a formula for its rate of radiation of energy.

16.4   Under the transformation of Problem 5.5, show that a static electric dipole of strength p_0 transforms into a magnetic dipole of strength $m_0 = cp_0$. What do you conjecture to be the formula for the rate of radiation of energy from a time-varying magnetic dipole?

16.5   A magnetic dipole of strength m_0 is rotating with angular frequency ω_0
        about a fixed axis inclined at an angle α to the dipole axis.
        Show that the rate of radiation of energy is

$$\frac{\mathrm{d}E}{\mathrm{d}t} = \frac{2}{3}\frac{\mu_0}{4\pi c^3} m_0^2 \omega_0^4 \sin^2 \alpha.$$

The earth has a dipole moment $m_0 \approx 8.0 \times 10^{22}$ A m^2 inclined at an angle
$\alpha \approx 11°$ to the rotation axis. Estimate the rate of change of the length of
day if this were due to the loss of rotational energy into magnetic dipole
radiation.
        (Moment of inertia of earth $\approx 8.1 \times 10^{37}$ kg m^2.)

17

Applications of radiation theory

17.1 Simple antennas

Hertz verified the existence of electromagnetic waves experimentally in 1887–8. Their potential utility in communications was soon recognised; Marconi began his successful series of experiments in long-distance 'wireless telegraphy' in 1894. The design of antennas for the transmission and reception of electromagnetic waves, from microwaves to radiowaves, is now a highly developed art.

We shall consider in detail only a simple antenna consisting of two thin straight wires of total length l, centre-fed at frequency ω, as in Fig. 17.1. If $\omega \ll c/l$, the current in the wires can be taken to be of the approximate form

$$I_0(1 - 2|z|/l)\cos \omega t = I(z)\cos \omega t, \tag{17.1}$$

choosing the z-axis along the wires. This expression is plausible (the current must vanish at the ends of the wire), and will be justified further in Chapter 18 (Problem (18.5)). Also, if $\omega \ll c/l$ the 'slowly varying' condition for (16.20) and (16.22) to hold is clearly met; hence the radiation is electric dipole and the radiation pattern is that given by Fig. 16.2. Using (16.21), the oscillating dipole moment $\mathbf{p} = (0, 0, p_z)$ is determined by

$$\frac{\mathrm{d}p_z}{\mathrm{d}t} = \int_{-l/2}^{l/2} I(z)\cos \omega t \,\mathrm{d}z = \tfrac{1}{2}I_0 l \cos \omega t,$$

so that $\ddot{p}_z = -\tfrac{1}{2}I_0 l\omega \sin \omega t$. \hfill (17.2)

From (16.26), the total rate of radiation of energy from the source through a sphere centred on the source, of radius $r \gg l$, at time t, is

$$\frac{\mathrm{d}\mathscr{E}}{\mathrm{d}t} = \frac{2}{3}\frac{1}{4\pi\varepsilon_0 c^3}\left(\frac{I_0 l\omega}{2}\right)^2 \sin^2 \omega(t - r/c).$$

144

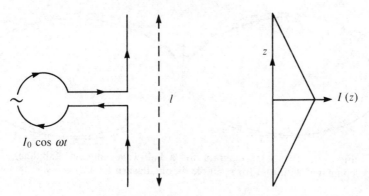

Fig. 17.1 A centre-fed antenna with $l \ll \lambda$ (left) and the current distribution $I(z)$ at $t = 0$ (right).

Averaging over many cycles, the mean rate is

$$\frac{\overline{\mathrm{d}\mathscr{E}}}{\mathrm{d}t} = \frac{l^2\omega^2 I_0^2}{48\pi\varepsilon_0 c^3} = \frac{\pi}{12}\left(\frac{l}{\lambda}\right)^2\left(\frac{\mu_0}{\varepsilon_0}\right)^{\frac{1}{2}} I_0^2, \qquad (17.3)$$

where $\lambda = 2\pi c/\omega$ is the wavelength of the radiation (and we have also used $c^2 = 1/\varepsilon_0\mu_0$). The condition $\omega \ll c/l$ is equivalent to $\lambda \gg l$.

$(\mu_0/\varepsilon_0)^{\frac{1}{2}} \approx 377 \; \Omega$ has the dimension of resistance, and we may write

▶
$$\frac{\overline{\mathrm{d}\mathscr{E}}}{\mathrm{d}t} = \tfrac{1}{2}R I_0^2, \qquad (17.4)$$

where R is called by convention the *radiation resistance* of the antenna.

For a given wavelength λ, and given I_0, an antenna with $l \ll \lambda$ is an inefficient way of radiating power since (17.3) contains the factor $(l/\lambda)^2$. In practice a *half-wave antenna* with $l = \lambda/2$ is used. The condition $l \ll \lambda$ no longer holds and we must use the more general expression (16.19). The current in a half-wave antenna is approximately the standing wave

$$I(z, t) = I_0 \cos(2\pi z/\lambda)\cos\omega t. \qquad (17.5)$$

Inserting this in (16.19) gives an expression for $\mathbf{A}(\mathbf{r}, t)$, which can be integrated in closed form (Problem 17.2), and hence expressions for $\mathbf{E}$ and $\mathbf{B}$. The resulting polar diagram for the radiated energy is qualitatively similar to that of a simple dipole but somewhat more directional (see Fig. 17.2). The mean rate of radiation of energy is

$$\frac{\overline{\mathrm{d}\mathscr{E}}}{\mathrm{d}t} = \tfrac{1}{2}R_{\lambda/2} I_0^2,$$

where the radiation resistance $R_{\lambda/2} = 73.1 \; \Omega$. This last result is obtained by a numerical integration (Problem 17.2).

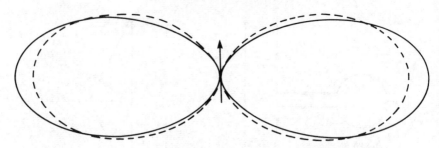

Fig. 17.2 The polar diagram for a half-wave antenna (full line) compared with that for a simple dipole (dashed line).

17.2 Arrays of antennas

Considerable directional control of the power distribution in radio transmission can be achieved by the judicious positioning of antennas in an array.

Consider first a single antenna at the origin, fed at frequency ω. We may take $\mathbf{J}(\mathbf{r}, t) = \mathbf{J}_0(\mathbf{r}) e^{i\omega t}$, using the complex form, so that (16.19) becomes

$$\mathbf{A}_0(\mathbf{r}, t) = \frac{\mu_0}{4\pi r} \exp\left[i\omega(t - r/c)\right] \int \mathbf{J}_0(\mathbf{r}') \exp\left(i\omega\hat{\mathbf{r}} \cdot \mathbf{r}'/c\right) dV'.$$

A similarly aligned antenna at a nearby point $\mathbf{R}$ will have a current distribution $\mathbf{J}_0(\mathbf{r} - \mathbf{R}) e^{i\omega t}$, and

$$\mathbf{A}_{\mathbf{R}}(\mathbf{r}, t) = \frac{\mu_0}{4\pi r} \exp\left[i\omega(t - r/c)\right] \int \mathbf{J}_0(\mathbf{r}' - \mathbf{R}) \exp\left(i\omega\hat{\mathbf{r}} \cdot \mathbf{r}'/c\right) dV'.$$

Making the change of variable $\mathbf{r}'' = \mathbf{r}' - \mathbf{R}$,

$$\mathbf{A}_{\mathbf{R}}(\mathbf{r}, t) = \frac{\mu_0}{4\pi r} \exp\left[i\omega(t - r/c + \hat{\mathbf{r}} \cdot \mathbf{R}/c)\right] \int \mathbf{J}_0(\mathbf{r}'') \exp\left(i\omega\hat{\mathbf{r}} \cdot \mathbf{r}''/c\right) dV''$$

$$= \exp\left[i(2\pi/\lambda)\hat{\mathbf{r}} \cdot \mathbf{R}\right] \mathbf{A}_0(\mathbf{r}, t).$$

Thus the field of the displaced antenna is simply related to that of the antenna at the origin by a phase factor. An array of N similar antennas at positions $\mathbf{R}_n$ oscillating in unison will give a resultant vector potential at a distant point $\mathbf{r}$

$$\mathbf{A}(\mathbf{r}, t) = \mathbf{A}_0(\mathbf{r}, t) \left\{ \sum_{n=1}^{N} \exp\left[i(2\pi/\lambda)\hat{\mathbf{r}} \cdot \mathbf{R}_n\right] \right\}.$$

More generally, if the antennas are fed with different phases ϕ_n

$$\mathbf{A}(\mathbf{r}, t) = \mathbf{A}_0(\mathbf{r}, t) \left\{ \sum_{n=1}^{N} \exp\left[i(2\pi/\lambda)\hat{\mathbf{r}} \cdot \mathbf{R}_n + i\phi_n\right] \right\}. \tag{17.6}$$

The phase factors in braces appear also in the expressions for the **B** and **E** fields. (The terms which come from differentiating $\hat{\mathbf{r}}$ are negligible for large r.) The expression for the time-averaged power radiated in the direction $\hat{\mathbf{r}}$ is easily seen to be

$$\blacktriangleright \qquad \mathbf{N} = \mathbf{N}_0 \left| \sum_{n=1}^{N} \exp\left[i(2\pi/\lambda) \hat{\mathbf{r}} \cdot \mathbf{R}_n + i\phi_n \right] \right|^2. \qquad (17.7)$$

Problems (17.4) and (17.5) illustrate how the choices of $\mathbf{R}_n$ and ϕ_n determine the directional characteristics of an array.

17.3 Radiation from a slowly moving charge

The general expression for $\mathbf{A}(\mathbf{r}, t)$ we obtained in Chapter 16 holds also on an atomic scale, and in particular it may be applied to the case of a slowly moving charged particle.

Consider a particle carrying charge Q, whose position is given by $\mathbf{r}_1(t)$, with its position at $t = 0$ taken as origin. If the velocity **v** of the particle is small compared with the velocity of light, we can use the approximations (16.20) and (16.22) for $\mathbf{A}(\mathbf{r}, t)$.

Since the dipole moment of the particle is

$$\mathbf{p}(t) = Q\mathbf{r}_1(t),$$

the radiated energy through a sphere of radius r at time t, using (16.26), is

$$\frac{d\mathcal{E}}{dt} = \frac{2}{3} \left(\frac{Q^2}{4\pi\varepsilon_0} \right) \frac{|\mathbf{a}(t - r/c)|^2}{c^3},$$

where $\mathbf{a} = \ddot{\mathbf{r}}_1$ is the acceleration of the particle. An important point which follows from this equation is that a charged particle moving with constant velocity does not radiate energy.

If the particle is accelerating, it radiates energy at a rate given by

$$\blacktriangleright \qquad \frac{d\mathcal{E}}{dt} = \frac{2}{3} \left(\frac{Q^2}{4\pi\varepsilon_0} \right) \frac{\mathbf{a}^2}{c^3}. \qquad (17.8)$$

This result is known as *Larmor's formula*.

17.4 The Abraham–Lorentz equation

Since energy is conserved, the energy lost into radiation when a particle is accelerating under given external forces, must come at the expense of the particle's kinetic energy: the particle must be subject to an additional damping force due to its own radiation. This *radiation damping* has not been included in our previous discussions of particle dynamics. It is fortunate that for many purposes the radiation damping can be considered to be a small perturbation (Problem 17.8).

The energy equation is an integral of the equations of motion. For velocities small compared with c, the Lorentz force law (4.1) can be modified in a way that will yield (17.8). Write

▶ $$m\ddot{\mathbf{r}} = Q(\mathbf{E} + \dot{\mathbf{r}} \times \mathbf{B}) + m\tau\dddot{\mathbf{r}}, \qquad (17.9)$$

where $\dddot{\mathbf{r}} = \mathrm{d}^3\mathbf{r}/\mathrm{d}t^3$.

Following the standard way of deriving the energy equation, we take the scalar product of (17.9) with $\dot{\mathbf{r}}$, and integrate with respect to t between t_1 and t_2; this gives

$$\frac{1}{2}m\dot{\mathbf{r}}^2 \bigg|_{t_1}^{t_2} = Q\int_{\mathbf{r}_1}^{\mathbf{r}_2} \mathbf{E} \cdot \mathrm{d}\mathbf{r} + m\tau\dot{\mathbf{r}}\cdot\ddot{\mathbf{r}}\bigg|_{t_1}^{t_2} - m\tau\int_{t_1}^{t_2} \ddot{\mathbf{r}}^2\,\mathrm{d}t. \qquad (17.10)$$

We have integrated $\dddot{\mathbf{r}}\cdot\dot{\mathbf{r}}$ by parts. In some circumstances the term $m\tau\dot{\mathbf{r}}\cdot\ddot{\mathbf{r}}|_{t_1}^{t_2}$ vanishes (for example, if the particle is moving with zero acceleration both initially and finally), and we then have the energy equation

$$\frac{1}{2}m\dot{\mathbf{r}}^2 \bigg|_{t_1}^{t_2} = Q\int_{\mathbf{r}_1}^{\mathbf{r}_2} \mathbf{E} \cdot \mathrm{d}\mathbf{r} - m\tau\int_{t_1}^{t_2} \ddot{\mathbf{r}}^2\,\mathrm{d}t. \qquad (17.11)$$

This is consistent with the expression (17.8) for the rate of radiation of energy if

$$\tau = \frac{2}{3}\left(\frac{Q^2}{4\pi\varepsilon_0}\right)\frac{1}{mc^3}. \qquad (17.12)$$

In the case of an electron, $\tau \approx 6.27 \times 10^{-24}$ s.

The equation of motion (17.9) is known as the *Abraham–Lorentz equation*. It has some peculiarities: it is a differential equation of third order in t, and possesses unphysical 'runaway' solutions (Problem 17.9), as well as solutions which are physically reasonable. We are again, as in our discussion of particle self energies in Chapter 2, meeting problems in classical electromagnetism which have not yet been wholly resolved.

17.5 The scattering of radiation by atoms and molecules

As an example of the use of Larmor's formula (17.8) we shall estimate the power radiated from an electron bound in an atom which is subject to an incident electromagnetic wave. In the simple atomic model of §12.3, we found, after equation (12.11), the induced dipole moment of the atom to be

$$\mathbf{p}(t) = \frac{(e^2/m_\mathrm{e})}{\omega_0^2 - \omega^2 - \mathrm{i}\gamma\omega}\mathbf{E}_0\,e^{-\mathrm{i}\omega t},$$

for incident wavelengths much larger than the size of the atom.

The real part of this gives

$$\mathbf{p}(t) = \frac{(e^2/m_e)}{[(\omega_0^2 - \omega^2)^2 + \gamma^2\omega^2]^{\frac{1}{2}}} \mathbf{E}_0 \cos(\omega t - \phi),$$

where $\phi = \tan^{-1}[\gamma\omega/(\omega_0^2 - \omega^2)]$.

Hence the atom radiates power at a mean rate

$$\frac{2}{3}\frac{1}{4\pi\varepsilon_0 c^3}\overline{|\ddot{\mathbf{p}}|^2} = \frac{4\pi}{3}\left(\frac{e^2}{4\pi\varepsilon_0}\right)^2 \frac{\varepsilon_0 E_0^2}{c^3 m_e^2}\frac{\omega^4}{[(\omega_0^2 - \omega^2)^2 + \gamma^2\omega^2]}. \qquad (17.13)$$

The radiated power must be taken from the incident wave, and the outgoing radiation can be regarded as the scattering of the incident radiation. It is useful to define the *scattering cross-section*

$$\sigma = \frac{\text{radiated power}}{\text{incident power/unit area}}.$$

Since the mean incident power per unit area is given by $c\varepsilon_0|\mathbf{E}_0|^2/2$ (equation (6.7)), we have for our model atom

$$\sigma = \sigma_T \frac{\omega^4}{[(\omega_0^2 - \omega^2)^2 + \gamma^2\omega^2]},$$

where

$$\sigma_T = \frac{8\pi}{3}\left(\frac{e^2}{4\pi\varepsilon_0 m_e c^2}\right)^2 \approx 0.665 \times 10^{-28} \text{ m}^2$$

is called the *Thomson scattering* cross-section: σ_T is the cross-section presented by a free electon (obtained by setting $\omega_0 = 0$ and $\gamma = 0$), and agrees well with the experimental cross-section at low frequencies ($\hbar\omega \ll m_e c^2$) where quantum effects are unimportant.

In a more realistic model of an atom or molecule, numerous natural frequencies ω_j, and damping constants γ_j, appear in $\mathbf{p}(t)$ (cf. the discussion in §12.4). However, in the case of many molecules there is a 'window', free from resonances, lying in the visible range of frequencies between the infrared molecular resonances and the ultraviolet electron resonances. Thus, in the visible range, the variation of scattering cross-section with frequency for these molecules is dominated by the ω^4 factor in (17.13): blue light is scattered more than red light. Given the energy spectrum of the radiation emitted from the sun, this explains the blue of the sky, and the red of the setting sun. Scattering dominated by an ω^4 dependence is known as *Rayleigh scattering*. It was first investigated by Lord Rayleigh in 1871.

Problems

17.1   Show that half the power from a vertical dipole antenna is radiated within about $\pm 20°$ of the horizontal.

17.2   Using (17.5) show that the vector potential of a half-wave antenna at a distant point is $(0, 0, A_z)$ where

$$A_z(\mathbf{r}, t) = \frac{\mu_0 I}{4\pi r} \int_{-\lambda/4}^{\lambda/4} \cos\left(\frac{2\pi z}{\lambda}\right) \cos \omega\left(t - \frac{r}{c} + \frac{z \cos \theta}{c}\right) dz$$

$$= \frac{\mu_0 I}{4\pi r}\left(\frac{\lambda}{2\pi}\right) \frac{2 \cos\left(\frac{1}{2}\pi \cos \theta\right)}{\sin^2 \theta} \cos \omega(t - r/c).$$

Hence show that the time-averaged rate of radiation of energy is

$$\frac{dE}{dt} = \frac{1}{4\pi}\left(\frac{\mu_0}{\varepsilon_0}\right)^{1/2} I_0^2 \int_0^\pi \frac{\cos^2\left(\frac{1}{2}\pi \cos \theta\right)}{\sin \theta} d\theta.$$

Compute the integral in this expression numerically, and show that the radiation resistance of a half-wave antenna is 73.1 Ω.

17.3   Estimate the current I_0 (equation (17.5)) needed to generate 1 kW of radiation power from a half-wave antenna.

For such a (vertical) antenna, find the maximum electric and magnetic field intensities at a point 10 km distant in the horizontal plane.

17.4   Television antennas sometimes consist of a number of vertical dipoles one above the other. This maintains a uniform power distribution around the axis of symmetry, but can produce a narrow horizontal distribution.

Show that two simple dipole antennas in phase so placed, a distance d apart, give a power distribution

$$P(\theta)\, d\theta = (\text{constant}) \sin^3 \theta \cos^2\left(\frac{\pi d \cos \theta}{\lambda}\right) d\theta$$

where λ is the wavelength and θ the angle with respect to the vertical.

What is a good value for d?

17.5   Find an expression for the pattern in the horizontal plane of the radiation emitted from two vertical half-wave antennas situated a quarter-wavelength apart in this plane and fed with the same signal but $\pi/2$ out of phase. Sketch the pattern.

17.6   A simple dipole antenna has radiation resistance R_1. Show that R_2, the radiation resistance of two such simple dipoles placed parallel and a short distance $D \ll \lambda$ apart, and oscillating completely out of phase, is approximately

$$R_2 \approx \frac{2}{5}\left(\frac{2\pi D}{\lambda}\right)^2 R_1.$$

Estimate R_2/R_1 for two power cables, 1 m apart and carrying a current at 50 Hz in opposite directions.

17.7   Consider classically the problem of an electron in a circular orbit of radius r about a proton.

Show that the total energy of the electron is

$$E = -\frac{1}{2}\left(\frac{e^2}{4\pi\varepsilon_0 r}\right).$$

The electron radiates according to Larmor's formula, and spirals towards the proton. Assume that at any time the orbit is approximately circular,

(a) estimate the time to fall from $r = 10\,\text{Å}$ to $r = 1\,\text{Å}$,

(b) show that classically the energy spectrum of the radiation is approximately

$$\frac{dE}{d\omega} = \frac{1}{3}\left(\frac{e^2}{4\pi\varepsilon_0}\right)^{2/3}\frac{m_e^{1/3}}{\omega^{1/3}}.$$

17.8   Consider the β-decay of fluorine to oxygen represented by the equation $^{18}_{9}\text{F} \to {}^{18}_{8}\text{O} + e^+ + \nu$.

Show that the total electromagnetic energy radiated by the positron after it has tunnelled through the Coulomb barrier is

$$E_\gamma \approx \frac{1}{12\sqrt{2}}E_e\left(\frac{E_e}{m_e c^2}\right)^{3/2}\int_1^\infty \frac{dx}{x^{7/2}(x-1)^{1/2}},$$

where E_e is the initial total (kinetic + potential) energy of the positron. Estimate the integral, and the energy radiated, if $E_e = 0.25$ MeV.

17.9   Show that in the absence of external forces the Abraham–Lorentz equation (17.9) has the 'runaway' solution $\ddot{\mathbf{r}}(t) = \ddot{\mathbf{r}}(0)\,e^{t/\tau}$, as well as the 'physical' solution $\ddot{\mathbf{r}} = 0$.

17.10   A bound electron oscillates at frequency ω_0. Show that the radiation damping of its motion may be represented approximately by a damping term $m_e\gamma\dot{\mathbf{r}}$ in the equation of motion, where

$$\gamma = \frac{2}{3}\left(\frac{e^2}{4\pi\varepsilon_0}\right)\frac{\omega_0^2}{m_e c^3}.$$

17.11   A dielectric sphere of radius a, real dielectric constant ε_r, is subject to a plane electromagnetic wave, wavelength $\lambda \gg a$. Show that the scattering cross-section of the sphere is

$$\sigma \approx \frac{8\pi a^2}{3}\left|\frac{\varepsilon_r - 1}{\varepsilon_r + 2}\right|^2\left(\frac{2\pi a}{\lambda}\right)^4.$$

(See §10.6).

18

Transmission lines, wave guides, and optical fibres

Transmission lines, wave guides, and optical fibres are basically channels for the transport of electromagnetic energy, for example communicating the pulsed signals between the various components of a computer network, or carrying the power from a radio-frequency generator to an antenna. In this chapter we shall outline the theory underlying their use.

18.1 Transmission lines

As a simple example of a transmission line, we shall consider a coaxial cable consisting of a central conducting wire of radius a, running along the z-axis, surrounded by a coaxial conducting cylinder of inner radius b (Fig. 18.1). The space between the conductors must, for mechanical reasons, be filled with a solid dielectric, but for simplicity we shall suppose for the moment this has $\varepsilon_r = \mu_r = 1$. We shall also suppose that the resistance of the conductors may be neglected. Then if a steady current I flows along the inner wire, and returns along the outer cylinder, the magnetic field lines will be circles around the wire with

$$\mathbf{B}(\mathbf{r}) = \frac{\mu_0 I}{2\pi}\left(-\frac{y}{\rho^2}, \frac{x}{\rho^2}, 0\right)$$

(cf. Problem 4.7).

The electric field between the wires will be radial, with

$$\mathbf{E}(\mathbf{r}) = \frac{V}{\ln(b/a)}\left(\frac{x}{\rho^2}, \frac{y}{\rho^2}, 0\right),$$

where V is the potential of the inner wire relative to the outer cylinder (which is usually grounded) (cf. Problem 7.1). Since we are neglecting the resistance of the conductors, there is no component of $\mathbf{E}$ in the z-direction.

If the fields change with time and with position along the cable, we may conjecture that I and V simply become functions of z and t:

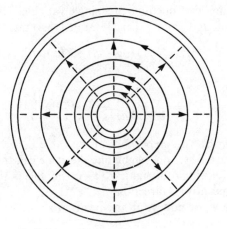

Fig. 18.1 A cross-section of a coaxial cable carrying a steady current. Field lines are shown for the **B** field (full lines) and **E** field (dashed lines), for the current in the inner conductor flowing out of the plane of the diagram, and the inner conductor at a higher potential than the outer.

$$\mathbf{B}(\mathbf{r}, t) = \frac{\mu_0 I(z, t)}{2\pi}\left(-\frac{y}{\rho^2}, \frac{x}{\rho^2}, 0\right),\qquad(18.1)$$

$$\mathbf{E}(\mathbf{r}, t) = \frac{V(z, t)}{\ln(b/a)}\left(\frac{x}{\rho^2}, \frac{y}{\rho^2}, 0\right).\qquad(18.2)$$

The forms of I and V are to be determined by Maxwell's equations in the region between the conductors. By inspection, our fields satisfy $\nabla \cdot \mathbf{B} = 0$, $\nabla \cdot \mathbf{E} = 0$. The Maxwell equation $\nabla \times \mathbf{E} + \partial \mathbf{B}/\partial t = 0$ is satisfied if

$$\frac{1}{\ln(b/a)}\frac{\partial V}{\partial z} = -\frac{\mu_0}{2\pi}\frac{\partial I}{\partial t},\qquad(18.3)$$

and the Maxwell equation $\nabla \times \mathbf{B} - (1/c^2)\partial \mathbf{E}/\partial t = 0$ is satisfied if

$$-\frac{\mu_0}{2\pi}\frac{\partial I}{\partial z} = \frac{1}{c^2\ln(b/a)}\frac{\partial V}{\partial t},\qquad(18.4)$$

as may readily be verified.

Taking $(\partial/\partial t)$ of (18.3) and $(\partial/\partial z)$ of (18.4), and eliminating $\partial^2 V/\partial z\,\partial t$, yields

$$\frac{\partial^2 I}{\partial z^2} = \frac{1}{c^2}\frac{\partial^2 I}{\partial t^2}.\qquad(18.5)$$

Thus $I(z, t)$ satisfies the one-dimensional wave equation, the general solution of which is

$$I(z, t) = I_+(z - ct) + I_-(z + ct),\qquad(18.6)$$

where I_+ and I_- are any functions of a single variable. The corresponding $V(z, t)$ is readily found from either (18.3) or (18.4) and is

▶ $$V(z, t) = Z_0[I_+(z - ct) - I_-(z + ct)], \qquad (18.7)$$

where

$$Z_0 = \frac{\mu_0 c}{2\pi} \ln \left(\frac{b}{a}\right) = \frac{1}{2\pi} \sqrt{\frac{\mu_0}{\varepsilon_0}} \ln \left(\frac{b}{a}\right)$$

is called the *characteristic impedance* of the cable; $\sqrt{(\mu_0/\varepsilon_0)} \approx 377 \, \Omega$.

Thus, if resistive losses are negligible, the cable can transmit any wave form with the velocity c of light, $I_+(z - ct)$ in the positive z-direction, $I_-(z + ct)$ in the negative z-direction.

The energy flow, or power, at any position z may be calculated using the Poynting vector (§6.5):

$$\text{power} = \frac{1}{\mu_0} \int \mathbf{E} \times \mathbf{B} \cdot d\mathbf{S},$$

where the integral is taken over the cross-section between the inner and outer conductors.

From (18.1) and (18.2),

$$\mathbf{E} \times \mathbf{B} = \frac{\mu_0}{2\pi \ln (b/a)} V(z, t) I(z, t) \left(0, 0, \frac{1}{\rho^2}\right),$$

so that

$$\begin{aligned}
\text{power} = P &= \frac{1}{2\pi \ln (b/a)} V(z, t) I(z, t) \int_a^b \frac{2\pi \rho \, d\rho}{\rho^2} \\
&= V(z, t) I(z, t) \\
&= Z_0[I_+^2(z - ct) - I_-^2(z + ct)], \qquad (18.8)
\end{aligned}$$

where the last line follows using (18.6) and (18.7). The energy travels with the wave forms. Curiously, the formulae are reminiscent of those of circuit theory, but (in our approximation) there are no resistive losses and there is no attenuation of the power.

We have so far set $\varepsilon_r = \mu_r = 1$. At sufficiently low frequencies, the dielectric between the conductors may usually be characterised by its static values of ε_r and μ_r. Including these factors in our equations, $\varepsilon_o \to \varepsilon_r \varepsilon_0$, $\mu_0 \to \mu_r \mu_0$, and $c \to c/\sqrt{(\varepsilon_r \mu_r)}$, so that $Z_0 \to Z_0 \sqrt{(\mu_r/\varepsilon_r)}$.

The coaxial cable is just one example of a transmission line. A simple pair of parallel wires, or parallel plates (Problem 18.4), will also serve as a transmission line. Again wave solutions of the Maxwell equations can be found in which, as in the coaxial cable, the $\mathbf{E}$ and $\mathbf{B}$ fields are transverse to the line and the waves propagate with velocity $c/\sqrt{(\varepsilon_r \mu_r)}$.

18.2 Attenuation due to resistive losses

In practice there are of course resistive losses in the conductors. We can estimate these losses as a perturbation on our 'ideal' solution. At high frequencies, when the skin depth is small compared with the radius of the inner conductor, we can treat the inner and outer conductor surfaces as planar, and apply the results of §13.3. For the materials used to make coaxial cables, $\mu_r = 1$ to a good approximation; the **B** field is therefore continuous across the dielectric–metal interfaces and (neglecting attenuation) parallel to them. Since we are no longer taking the conductors to be perfect, an electric field will exist in them in the direction of current flow, along the cable and, hence, perpendicular to the **B** field. Thus in both inner and outer conductors a plane wave attenuating away from the surface is set up. In the metal, **E** and **B** are related as in equation (13.15), with $|\mathbf{E}| = c|\mathbf{B}|/\sqrt{2n}$. The time-averaged Poynting vector which gives the flow of energy per unit area into the metal is (using (12.22) with $z = 0$)

$$|\mathbf{N}| = n(E_0^2/2\mu_0 c) = cB_0^2/4\mu_0 n.$$

From (13.9), $n = \sqrt{(\sigma_0/2\varepsilon_0 \omega)}$, so that

$$|\mathbf{N}| = \left(\frac{2\omega}{\mu_0 \sigma_0}\right)^{\frac{1}{2}} \frac{B_0^2}{4\mu_0}, \tag{18.9}$$

where σ_0 is the conductivity of the conductors.

Considering a travelling wave at frequency ω, with $I = I_0 \cos(kz - \omega t)$, where $k = \omega\sqrt{(\varepsilon_r \mu_r)}/c$, the time-averaged power loss into an element $\mathrm{d}z$ of the cable is

$$-\mathrm{d}P = \frac{1}{4\mu_0}\left(\frac{2\omega}{\mu_0 \sigma_0}\right)^{\frac{1}{2}}\left(\frac{\mu_0 I_0}{2\pi}\right)^2\left(\frac{2\pi a}{a^2}+\frac{2\pi b}{b^2}\right)\mathrm{d}z,$$

where we have used (18.1) for the fields at the surface of the conductors, and added the contributions from the inner and outer conductors.

Hence, with the help of (18.8),

$$\frac{\mathrm{d}P}{\mathrm{d}z} = -\frac{P}{d}, \tag{18.10}$$

where the *attenuation length d* is given by

$$d = \left(\frac{2\sigma_0}{\omega\varepsilon_0}\right)^{\frac{1}{2}} \frac{ab}{(a+b)}\left(\frac{1}{\varepsilon_r}\right)^{\frac{1}{2}} \ln\left(\frac{b}{a}\right). \tag{18.11}$$

From (18.10), the power decays exponentially along the cable. From (18.11) the attenuation length decreases as the frequency increases. It is this effect, along with losses due to absorption in the dielectric at higher

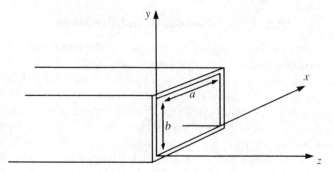

Fig. 18.2 Coordinate axes taken for the rectangular wave guide.

frequencies, which limits the use of coaxial cable to low frequency transmission. At microwave frequencies, the *wave guide* becomes more suitable for the transmission of power.

18.3 Wave guides

A wave guide is a hollow metallic pipe, usually of rectangular cross-section. There is no inner conductor. We shall set $\varepsilon_r = 1$, $\mu_r = 1$ for the ambient air which normally will fill the pipe, and suppose that we may treat the walls of the guide as perfectly conducting.

The wave forms in a guide are more complicated than those used in a coaxial cable, and for this reason we shall treat only waves at a fixed frequency ω. We use the complex form $\exp[i(kz - \omega t)]$ for mathematical convenience.

Consider a rectangular wave guide of inner dimensions a, b, lying along the z-axis (Fig. 18.2). For perfectly conducting walls, the boundary conditions are

$$E_y = E_z = 0 \text{ for } x = 0 \text{ and } x = a,$$
$$E_x = E_z = 0 \text{ for } y = 0 \text{ and } y = b.$$

A simple wave form which satisfies these conditions is

$$\mathbf{E} = \left[\alpha \cos\left(\frac{m\pi x}{a}\right) \sin\left(\frac{n\pi y}{b}\right), \beta \sin\left(\frac{m\pi x}{a}\right) \cos\left(\frac{n\pi y}{b}\right), \right.$$

$$\left. \gamma \sin\left(\frac{m\pi x}{a}\right) \sin\left(\frac{n\pi y}{b}\right) \right] e^{i(kz-\omega t)}, \quad (18.12)$$

provided that both n and m are integers. α, β, and γ are complex numbers.

So much for the boundary conditions! To satisfy the Maxwell equation $\nabla \cdot \mathbf{E} = 0$ we need

$$-\frac{m\pi}{a}\alpha - \frac{n\pi}{b}\beta + ik\gamma = 0.$$

This eliminates one parameter, say

$$\gamma = -\frac{i\pi}{k}\left(\frac{m\alpha}{a} + \frac{n\beta}{b}\right). \tag{18.13}$$

The magnetic field can be determined from the Maxwell equation $\nabla \times \mathbf{E} + \partial\mathbf{B}/\partial t = 0$, which gives

$$\mathbf{B} = \frac{1}{i\omega}\left[\left(\frac{n\pi\gamma}{b} - ik\beta\right)\sin\left(\frac{m\pi x}{a}\right)\cos\left(\frac{n\pi y}{b}\right),\right.$$

$$\left(ik\alpha - \frac{m\pi\gamma}{a}\right)\cos\left(\frac{m\pi x}{a}\right)\sin\left(\frac{n\pi y}{b}\right),$$

$$\left.\left(\frac{m\pi\beta}{a} - \frac{n\pi\alpha}{b}\right)\cos\left(\frac{m\pi x}{a}\right)\cos\left(\frac{n\pi y}{b}\right)\right]e^{i(kz-\omega t)}. \tag{18.14}$$

The Maxwell equation $\nabla \cdot \mathbf{B} = 0$ is then satisfied identically. Each component of the remaining Maxwell equation, $\nabla \times \mathbf{B} - (1/c^2)\partial\mathbf{E}/\partial t = 0$, yields the same condition

▶
$$\frac{\omega^2}{c^2} = \frac{m^2\pi^2}{a^2} + \frac{n^2\pi^2}{b^2} + k^2. \tag{18.15}$$

This *dispersion relation* is obtained more directly by noting that the components of $\mathbf{E}$ and $\mathbf{B}$ in free space satisfy the wave equation. The phase velocity (ω/k) is greater than the velocity of light, but not, of course, the group velocity $d\omega/dk$ (Problem 18.6). It should be noted that dispersion cannot be avoided when signals are propagated down a wave guide.

18.4 Modes of propagation

For a propagating solution k must be real, and it follows from (18.15) that for any pair of integers m, n there is a minimum frequency of propagation ω_{mn}, where

$$\frac{\omega_{mn}^2}{c^2} = \frac{m^2\pi^2}{a^2} + \frac{n^2\pi^2}{b^2}.$$

If neither m nor n is zero, then for $\omega > \omega_{mn}$ there are two linearly independent solutions of the Maxwell equations, corresponding to the two remaining parameters α, β. We may choose these to be the *transverse electric*, TE mode, with $E_z = 0$ everywhere (taking $(m\alpha/a) + (n\beta/b) = 0$) and *transverse magnetic*, TM mode, with $B_z = 0$ everywhere (taking $(m\beta/a) - (n\alpha/b) = 0$).

If $a > b$, the lowest cut-off frequency is given by $m = 1, n = 0$. Only the TE$_{1,0}$ mode exists (see (18.12)) and

$$\frac{\omega_{10}}{c} = \frac{\pi}{a}.$$

This is the mode usually used in practice. The cut-off frequency corresponds to a free space wavelength of $2a$. Thus if a is a few centimetres, the lowest propagating frequencies are in the microwave range.

For frequencies between ω_{10} and that of the next mode (ω_{20} or ω_{01}, depending on whether $(a/b) > 2$ or $(a/b) < 2$) only the $TE_{1,0}$ mode propagates. A wave guide is usually operated within this range; unwanted complications arise if there is more than one mode with the same frequency, since such modes would be mixed at, say, a bend in the wave guide, and then travel with different velocities (since they would have different wave numbers).

Setting $m = 1$, $n = 0$ in (18.12) and taking the real part, the electric field in the $TE_{1,0}$ mode is

$$\mathbf{E} = \left[0, E_0 \sin\left(\frac{\pi x}{a}\right) \cos(kz - \omega t), 0 \right], \qquad (18.16)$$

writing E_0 for the parameter β. E_0 is the maximum value of the electric field inside the guide. The magnetic field is then given by (18.14), with $\alpha = \gamma = 0$:

$$\mathbf{B} = \left[-\frac{kE_0}{\omega} \sin\left(\frac{\pi x}{a}\right) \cos(kz - \omega t), 0, \frac{\pi E_0}{a\omega} \cos\left(\frac{\pi x}{a}\right) \sin(kz - \omega t) \right]. \qquad (18.17)$$

The energy flow in the guide is obtained by constructing the Poynting vector from (18.15) and (18.16). The time-averaged Poynting vector is in the z-direction, and integrated over the cross-section of the guide gives the power transmitted, $E_0^2 kab/4\mu_0 \omega$ (Problem 18.7).

So far our calculations have assumed that the walls of the guide have zero resistivity. The losses in a real guide can be estimated using the formula (18.9), as in the calculation for a coaxial cable in §18.2. Typical attenuation lengths for copper guides at microwave frequencies are $\sim 10^2$ m (Problem 18.8).

18.5 Optical fibres

Dielectrics can also be used to guide waves. An *optical fibre* consists of an inner cylindrical core, usually a glass of high purity, surrounded by a cladding of a lower dielectric constant. The fibre is enclosed in a further jacket to provide protection. Such a system can support an electromagnetic wave, in rather the same way as a metallic wave guide, though the boundary conditions to be satisfied by the solutions of the Maxwell equations are of course different. The cladding carries an evanescent wave.

We saw in §12.6 that many dielectric materials have a 'transparency window' around the visible region of the electromagnetic spectrum, where

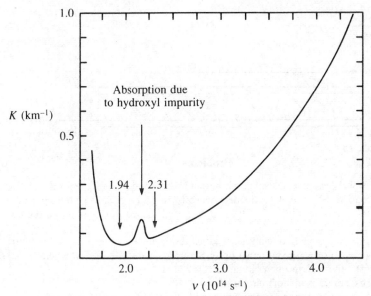

Fig. 18.3 Absorption coefficient K for a typical optical fibre as a function of frequency v. Standard operating frequencies are indicated. The rapid increase in K at the higher frequencies is due to Rayleigh scattering.

absorption is low. It is at frequencies in such a window (corresponding to free space wavelengths in the range $0.8\,\mu\text{m} < \lambda < 1.6\,\mu\text{m}$, in the near infrared) that optical fibres are used (Fig. 18.3).

If the core diameter is of the order of a few free space wavelengths, 'single mode' operation is possible (cf. §18.4). Single mode fibres are increasingly used for high quality, long distance transmission, over distances of the order of tens (or even hundreds) of kilometres.

In less demanding applications, thicker fibres with core diameters up to $200\,\mu\text{m}$ are used. Then many modes will be excited at the frequency of operation, and an analysis in terms of solutions of the Maxwell equations is less helpful. It is more convenient to take geometrical optics as a starting point, and describe the transmission of light as the repeated internal reflection of rays at the boundary between the core and the cladding.

In single mode fibres the cladding must be of much greater diameter than the core, and may carry $\sim 70\,\%$ of the energy. In multimode fibres, most of the energy is confined to the core. In either case the fibres are exceedingly fine; the standard cladding diameter is $125\,\mu\text{m}$. The technology is of great interest and importance.

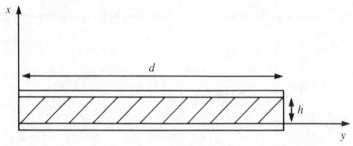

Fig. 18.4

Problems

18.1   A 'lossless' coaxial cable of characteristic impedance Z_0 feeds a resistive load R. Show that an incident wave form $I_i(ct-z)$ is in general partially reflected, but if $R = Z_0$ all the power is absorbed in the load and there is no reflected wave.

18.2   For waves at frequency ω, what is the input impedance $Z_i = V(z = 0)/I(z = 0)$ of a coaxial cable of length $\lambda/4$, where $\lambda = 2\pi c/\omega$, when
(a)   there is an open circuit at $z = \lambda/4$,
(b)   there is a short circuit at $z = \lambda/4$?
Show that if there is a resistive load Z, then $Z_i = Z_0^2/Z$.

18.3   Show that for a coaxial cable of fixed outer radius b, the attenuation is minimised if the inner radius $a = 0.28b$.

18.4   Consider two long parallel conducting strips, width d, distance h apart, where $h \ll d$ (see Fig. 18.4). The space between them is filled with dielectric of permittivity ε_r (and $\mu_r = 1$). A steady current I flows down one strip and returns along the other.
(a)   Sketch the magnetic field lines between the strips.
(b)   Sketch the electric field lines when one conductor is maintained at potential V relative to the other.
(c)   Neglecting edge effects, what are the magnitudes of these fields?
Suppose the strips are operated as a transmission line. Construct appropriate trial solutions of Maxwell's equations in the region between the strips, neglecting edge effects and losses in the conductors. Hence show that the transmission line has a characteristic impedance

$$Z_0 = \left(\frac{\mu_0}{\varepsilon_0 \varepsilon_r}\right)^{\frac{1}{2}} \frac{h}{d}.$$

Such strips are used in printed circuits at GHz frequencies. What might be their advantages and disadvantages?

18.5   Show that a transmission line of characteristic impedance Z_0, terminated at $z = 0$ as an open circuit, can support standing waves for $z \leqslant 0$ of the form

$$I = A \sin\frac{2\pi z}{\lambda} \cos\frac{2\pi ct}{\lambda},$$

$$V = -Z_0 A \cos\frac{2\pi z}{\lambda} \sin\frac{2\pi ct}{\lambda},$$

where A is a constant.

A simple dipole antenna of length $l \ll \lambda$ (§17.1) is like a short length of transmission line. Show that the current will be approximately a standing wave of the form of equation (17.1) with $I_0 = -\pi A l/\lambda$, and the impedance of the antenna will be imaginary, of magnitude $Z_0(\lambda/\pi l)$.

18.6 Show that the phase velocity v_{phase} and group velocity v_{group} of waves in a rectangular wave guide are related by

$$v_{group} = c^2/v_{phase}.$$

What is the significance of this result?

18.7 Show that the power transmitted by a wave guide in the $TE_{1,0}$ mode is $E_0^2 kab/4\mu_0 \omega$.

A rectangular wave guide transmits 3 GHz power to a radar antenna. The cross-section of the guide is 6 cm × 4 cm, and the electric field in the guide must not exceed $20 \, kV \, cm^{-1}$ to avoid the risk of dielectric breakdown of the air. Find the maximum power that may be transmitted.

What is the lowest frequency that can be propagated without attenuation down this guide?

If an attempt is made to propagate a wave at 0.99 times this frequency, at what distance is the power reduced by a factor of $1/e$?

For what range of frequency is the guide most suitable?

18.8 *Attenuation of the $TE_{1,0}$ mode in a wave guide.* Show that the power loss is given by

$$-\frac{dP}{dz} = \frac{1}{4\mu_0}\left(\frac{2\omega}{\mu_0 \sigma}\right)^{\frac{1}{2}}\left(\frac{\pi E_0}{a\omega}\right)^2 2b \; \text{(from } x = 0, x = a, \text{ faces)},$$

$$+\frac{1}{4\mu_0}\left(\frac{2\omega}{\mu_0 \sigma}\right)^{\frac{1}{2}} E_0^2 \left[\left(\frac{\pi}{a\omega}\right)^2 + \frac{k^2}{\omega^2}\right] a \; \text{(from } y = 0, y = b, \text{ faces)},$$

and hence derive the formula for the attenuation length D:

$$-\frac{(dP/dz)}{P} = \frac{1}{D} = \frac{1}{c}\left(\frac{2\omega}{\mu_0 \sigma}\right)^{\frac{1}{2}} \frac{\left[\dfrac{1}{b} + \dfrac{2}{a}\left(\dfrac{\omega_{10}}{\omega}\right)^2\right]}{\left[1 - \left(\dfrac{\omega_{10}}{\omega}\right)^2\right]^{\frac{1}{2}}}.$$

(Note that $\omega_{10}/c = \pi/a$, $k^2 + (\pi^2/a^2) = \omega^2/c^2$.)

Sketch D as a function of (ω/ω_{10}). Estimate D for Problem 18.7 if the wave guide is made of copper ($\sigma = 6.5 \times 10^7 \, \Omega^{-1} \, m^{-1}$).

19

The electromagnetic field and special relativity

19.1 The principle of relativity

Underlying all our discussion of electric and magnetic fields has been a coordinate system to indicate positions (x, y, z) in space, and clocks to indicate the time t at those positions. Maxwell's equations were abstracted from laboratory experiments, but implicitly our space–time reference frame has been an 'inertial frame', which is defined as one in which a freely moving body moves without acceleration. An earth-bound laboratory approximates to an inertial frame if a gravitational field is introduced to describe the effect of the earth's mass. In writing down the Lorentz force equation (4.1), we omitted the gravitational term. The reader should insert it when it is needed to describe a particular situation.

According to the principle of relativity, the laws of physics are the same in all inertial frames of reference. A triumph of the Maxwell theory is the interpretation of light waves as electromagnetic radiation, with velocity $c = 1/\sqrt{(\varepsilon_0 \mu_0)}$ in empty space. Thus if Maxwell's equations have universal validity, the equations have the same form, and the velocity of light has the same value, in any inertial frame of reference.

19.2 The Lorentz transformation

In introductory texts on the special theory of relativity, the Lorentz transformation, which relates events in two inertial frames of reference S and S', is derived from the postulate that the velocity of light is the same when measured in any inertial frame of reference. Suppose that S' is moving with velocity u along the x-axis of S and the origins and axes of S and S' coincide at $t = t' = 0$. Consider an event which occurs at (x, y, z, t) in S, and (x', y', z', t') in S'. It is found that the coordinates and times are related by

$$x' = \gamma(x - ut), \quad y' = y, \quad z' = z, \quad t' = \gamma(t - ux/c^2), \qquad (19.1)$$

where $\gamma = (1 - u^2/c^2)^{-\frac{1}{2}}$, and c is the velocity of light.

The Lorentz transformation (19.1) was obtained from this postulate in Einstein's famous 1905 paper. However, the transformation had been found previously by Lorentz and by other workers who had looked for a mathematical transformation which retained the form of Maxwell's equations (or the wave equation) in going from one coordinate system to another. The corresponding Galilean transformation of Newtonian physics ($x' = x - ut$, $y' = y$, $z' = z$, $t' = t$) does not have this property.

To apply the Lorentz transformation to Maxwell's equations is straightforward, if rather lengthy. Using the chain rule gives, for example

$$\frac{\partial E_x}{\partial x} = \frac{\partial E_x}{\partial x'}\frac{\partial x'}{\partial x} + \frac{\partial E_x}{\partial t'}\frac{\partial t'}{\partial x}$$

$$= \gamma\frac{\partial E_x}{\partial x'} - \frac{\gamma u}{c^2}\frac{\partial E_x}{\partial t'}.$$

Maxwell's equations are recovered in the primed frame provided one identifies

$$E'_x = E_x, \quad E'_y = \gamma(E_y - uB_z), \quad E'_z = \gamma(E_z + uB_y);$$

$$B'_x = B_x, \quad B'_y = \gamma\left(B_y + \frac{u}{c^2}E_z\right), \quad B'_z = \gamma\left(B_z - \frac{u}{c^2}E_y\right); \tag{19.2}$$

$$J'_x = \gamma(J_x - u\rho), \quad J'_y = J_y, \quad J'_z = J_z, \quad \rho' = \gamma\left(\rho - \frac{u}{c^2}J_x\right). \tag{19.3}$$

These results appear in the latter part of Einstein's 1905 paper.

Finally, the Lorentz force equation

$$\frac{d\mathbf{p}}{dt} = Q(\mathbf{E} + \mathbf{v} \times \mathbf{B}), \tag{19.4}$$

together with the energy equation

$$\frac{d\mathscr{E}}{dt} = Q\mathbf{E}\cdot\mathbf{v}, \tag{19.5}$$

transform into

$$\frac{d\mathbf{p}'}{dt'} = Q(\mathbf{E}' + \mathbf{v}' \times \mathbf{B}'), \quad \frac{d\mathscr{E}'}{dt'} = Q\mathbf{E}'\cdot\mathbf{v}',$$

provided that $(\mathscr{E}/c, \mathbf{p})$ is taken to be the relativistic energy–momentum four-vector with $\mathscr{E} = \gamma mc^2$, $\mathbf{p} = \gamma m\mathbf{v}$, and the charge Q is invariant under a Lorentz transformation (Problem 19.3).

We see from (19.2) that the electric and magnetic fields are inextricably linked. For example, what appears as a purely electrostatic field in one frame of reference will in any other frame manifest itself as a combination of electric and magnetic fields.

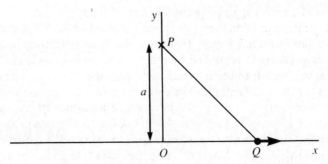

Fig. 19.1 The charge Q is moving with velocity u along the x-axis.

19.3 The fields of a charge moving with uniform velocity

As an application of the transformation equations, we shall calculate the electric and magnetic fields due to a particle carrying charge Q moving with uniform velocity in a frame S. We take the charge to be moving along the x-axis with velocity u, so that its position is given by $x = ut$. In the frame S', with the particle at rest at the origin, $\mathbf{B}' = 0$ and $\mathbf{E}'$ is given by Coulomb's law: $\mathbf{E}' = (Q/4\pi\varepsilon_0)\mathbf{r}'/r'^3$.

The fields in S are given by interchanging primed and unprimed coordinates and writing $-u$ for u in (19.2), so that in terms of the primed coordinates:

$$E_x = E'_x = \frac{Qx'}{4\pi\varepsilon_0 r'^3}, \quad E_y = \gamma E'_y = \frac{\gamma Qy'}{4\pi\varepsilon_0 r'^3}, \quad E_z = \gamma E'_z = \frac{\gamma Qz'}{4\pi\varepsilon_0 r'^3};$$

$$B_x = 0, \quad B_y = -\frac{\gamma u}{c^2}E_z = -\frac{\mu_0}{4\pi}\frac{Q\gamma uz'}{r'^3}, \quad B_z = \frac{\gamma u}{c^2}E'_y = \frac{\mu_0}{4\pi}\frac{Q\gamma uy'}{r'^3}.$$

(19.6)

Let us calculate the field at the point P, $(0, a, 0)$, at time t (Fig. 19.1). Then, in S', $x' = \gamma(0 - ut) = -\gamma ut$, $y' = y = a$, $z' = z = 0$, and $r' = (a^2 + \gamma^2 u^2 t^2)^{\frac{1}{2}}$.

Thus, from (19.6), the magnetic field at P is

$$B_x = 0, \quad B_y = 0, \quad B_z = \frac{\mu_0}{4\pi}\frac{Q\gamma ua}{(a^2 + \gamma^2 u^2 t^2)^{3/2}}.$$

(19.7)

In the low velocity limit, this becomes

$$B_x = 0, \quad B_y = 0, \quad B_z = \frac{\mu_0}{4\pi}\frac{Qua}{r^3},$$

(19.8)

which is just what we would have found by assuming that the Biot–Savart law (9.9) held for a single charged particle. It is an interesting exercise to investigate how the exact result (19.7), applied to electrons flowing in a linear wire to make up a current I, yields the Biot–Savart result *without* the approximation $u \ll c$, as must be the case since the Biot–Savart law follows from the Maxwell equations! (Problem 19.5).

The electric field at P is

$$E_x = -\frac{(Q/4\pi\varepsilon_0)\gamma ut}{[a^2 + \gamma^2 u^2 t^2]^{3/2}}, \quad E_y = \frac{(Q/4\pi\varepsilon_0)\gamma a}{[a^2 + \gamma^2 u^2 t^2]^{3/2}}, \quad E_z = 0. \quad (19.9)$$

For large velocities $u \lesssim c$, γ becomes large, and the electric field at P is only significant for $t \approx 0$, over a small time interval $\Delta t \sim a/\gamma u$ (Problem 19.4).

19.4 Four-vectors

In discussing relativity, it is useful to introduce the notion of a 'four-vector'. As an example, a point $\mathbf{r}$ in space at a time t can be thought of as a vector in the four dimensions of space–time. To have all the components of the vector with the same physical dimension, we multiply the time t by c, and denote such a point by $(ct, \mathbf{r})$. The Lorentz transformation (19.1) is then more symmetrical:

$$x' = \gamma[x - (u/c)\,ct], \quad y' = y, \quad z' = z, \quad ct' = \gamma[ct - (u/c)x]. \quad (19.10)$$

More generally, any set of four quantities which transform in this way is called a four-vector. We see from (19.3) that at every point in space and time the quantities $(c\rho, \mathbf{J})$ make up a four-vector, so that $(c\rho, \mathbf{J})$ is a four-vector field. Also we saw in §16.1 that in the gauge in which

$$\nabla \cdot \mathbf{A} + \frac{1}{c^2}\frac{\partial \Phi}{\partial t} = 0 \quad (19.11)$$

the potentials $\mathbf{A}$ and Φ satisfy

$$\nabla^2 \Phi - \frac{1}{c^2}\frac{\partial^2 \Phi}{\partial t^2} = -\frac{\rho}{\varepsilon_0}, \quad \nabla^2 \mathbf{A} - \frac{1}{c^2}\frac{\partial^2 \mathbf{A}}{\partial t^2} = -\mu_0 \mathbf{J}.$$

We can write these equations together as

$$\left(\nabla^2 - \frac{1}{c^2}\frac{\partial^2}{\partial t^2}\right)(\Phi/c, \mathbf{A}) = -\mu_0(c\rho, \mathbf{J}), \quad (19.12)$$

since $c^2 = 1/\mu_0\varepsilon_0$.

One may verify, again using the chain rule, that under a Lorentz transformation

$$\nabla^2 - \frac{1}{c^2}\frac{\partial^2}{\partial t^2} = \nabla'^2 - \frac{1}{c^2}\frac{\partial^2}{\partial t'^2}. \quad (19.13)$$

Since (19.12) should have the same form in every inertial frame, $(\Phi/c, \mathbf{A})$ must transform like $(c\rho, \mathbf{J})$ and be a four-vector field also.

As for the $\mathbf{E}$ and $\mathbf{B}$ fields themselves, the mathematical structure of the transformations (19.2) is not immediately clear. In a more abstract formulation, it appears that the components of $\mathbf{E}$ and $\mathbf{B}$ make up an antisymmetric 'four-tensor', of the form

$$\begin{pmatrix} 0 & -E_x & -E_y & -E_z \\ E_x & 0 & -cB_z & cB_y \\ E_y & cB_z & 0 & -cB_x \\ E_z & -cB_y & cB_x & 0 \end{pmatrix}.$$

The transformations (19.2) then follow from the transformation rules for four-tensors.

In the preceding chapters we have regarded the scalar potential Φ and the vector potential $\mathbf{A}$ as mathematical devices. However, although they are defined only up to a gauge transformation, the potentials contain within them all six components of the electric and magnetic fields. The emphasis in this book has been on the electric and magnetic fields, in free space and in matter. At the microscopic level, however, it is the potentials which take on a central role in the further development of the theory: the quantisation of the electromagnetic field to reveal the photon, and the fundamental theory of the interactions between particles and fields.

Problems

19.1   In non-relativistic mechanics, a Galilean transformation gives $\mathrm{d}\mathbf{p}'/\mathrm{d}t' = \mathrm{d}\mathbf{p}/\mathrm{d}t$; $\mathbf{v}' = \mathbf{v} - \mathbf{u}$. Use these equations, with the Lorentz force law, to obtain the transformation $\mathbf{E}' = \mathbf{E} + \mathbf{u} \times \mathbf{B}$, $\mathbf{B}' = \mathbf{B}$, which is the limiting form of (19.2) when $u \ll c$, and $|\mathbf{E}| \ll c|\mathbf{B}|$.

19.2   Show that in the relativistic generalisation of Problem 4.1 the cyclotron frequency becomes $\omega = eBc^2/\mathscr{E}$.

19.3   Verify the transformation properties of equations (19.4) and (19.5). (Multiply through first by $\gamma = \mathrm{d}t/\mathrm{d}\tau$, where τ is the invariant proper time.)

19.4   From equations (19.7), sketch the forms of the components of the electric field at P as functions of time for (a) $\gamma \simeq 1$, (b) $\gamma \gg 1$.

19.5   Suppose an electric current I flowing in a long straight wire along the z-axis to consist of charges ΔQ moving with velocity u, which at some instant $t = 0$ are at the points $x_n = nu\Delta t$, so that $I = \Delta Q/\Delta t$. Starting from equations (19.7), show that in the limit $\Delta Q \to 0$, $\Delta t \to 0$, the field at a point $(0, a, 0)$ is $(0, 0, B_z)$, where $B_z = \mu_0 I/(2\pi a)$.

19.6   Under a gauge transformation,

$$\Phi \to \Phi - \frac{\partial\chi}{\partial t}, \quad \mathbf{A} \to \mathbf{A} + \nabla\chi.$$

Show that $[-(1/c)\partial\chi/\partial t, \nabla\chi]$ is a four-vector, provided that χ is a scalar field: $\chi(\mathbf{r}', t') = \chi(\mathbf{r}, t)$.

19.7   Verify that the gauge condition (19.11) transforms into $\nabla' \cdot \mathbf{A}' + (1/c^2)\partial\Phi'/\partial t' = 0$.

Appendix A

Proof of Gauss's theorem

The electric field due to a point charge Q at the origin is $\mathbf{E(r)} = Q\hat{\mathbf{r}}/4\pi\varepsilon_0 r^2$. Hence the electric flux through an arbitrary closed surface S is

$$\int_S \mathbf{E} \cdot d\mathbf{S} = \frac{Q}{4\pi\varepsilon_0} \int_S \frac{\hat{\mathbf{r}} \cdot d\mathbf{S}}{r^2}. \tag{A.1}$$

Now $\hat{\mathbf{r}} \cdot d\mathbf{S} = \hat{\mathbf{r}} \cdot \hat{\mathbf{n}} \, dS$ is the projection of $d\mathbf{S}$ normal to $\mathbf{r}$. Divided by r^2, it is dimensionless, and numerically equal to $\pm d\Omega$, where $d\Omega$ is the area cut out on a sphere of unit radius by the cone whose apex is the origin and base is dS, as in Fig. A.1. The sign $\pm$ depends on the direction of the normal $\hat{\mathbf{n}}$ relative to $\mathbf{r}$. ($d\Omega$ is called the *solid angle* subtended at O by the surface element dS.)

If the surface S encloses the origin (Fig. A.2),

$$\int \frac{\hat{\mathbf{r}} \cdot d\mathbf{S}}{r^2} = 4\pi,$$

since the area of the unit sphere is 4π, and $\hat{\mathbf{n}}$ is taken to be the outward

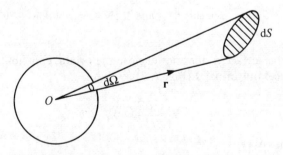

Fig. A.1 The solid angle $d\Omega$ subtended by an element of area dS.

167

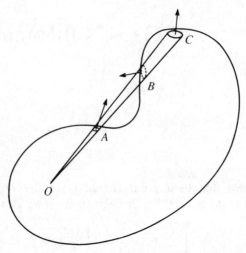

Fig. A.2 Surface elements at *A*, *B*, *C* give contributions $+d\Omega$, $-d\Omega$, and $+d\Omega$ to the surface integral.

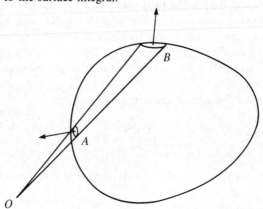

Fig. A.3 Surface elements at *A* and *B* give contributions $-d\Omega$ and $+d\Omega$.

normal. If the surface *S* does not enclose the origin, contributions to the integral cancel out in pairs (Fig. A.3) so that

$$\int \frac{\hat{\mathbf{r}} \cdot d\mathbf{S}}{r^2} = 0.$$

Hence, from (A.1)

▶ $$\int_S \mathbf{E} \cdot d\mathbf{S} = \begin{cases} Q/\varepsilon_0 & \text{if } Q \text{ lies inside } S, \\ 0 & \text{if } Q \text{ lies outside } S. \end{cases} \qquad (\text{A.2})$$

This result is known as Gauss's theorem.

Appendix B

The uniqueness theorem

Suppose we seek solutions of Laplace's equation, $\nabla^2\Phi = 0$, in the region of space V between finite conducting bodies with surfaces $S_1, S_2, \ldots$. What conditions suffice to define the solution uniquely? If V extends to infinity, we enclose the system in a sphere S_R of radius R, and consider the limit $R \to \infty$.

Let $\Phi_1(\mathbf{r})$, $\Phi_2(\mathbf{r})$ be two, possibly different, solutions. We construct the function $f(\mathbf{r}) = \Phi_1(\mathbf{r}) - \Phi_2(\mathbf{r})$. Since $\nabla^2\Phi_1 = 0$ and $\nabla^2\Phi_2 = 0$, then $\nabla^2 f = 0$ also.

Now $\int_V |\nabla f|^2 \, dV \geqslant 0$, since the integrand is positive or zero; the equality holds only if $\nabla f = 0$ everywhere in V.

Using the vector identity $\nabla \cdot (f\nabla f) = \nabla f \cdot \nabla f + f\nabla^2 f$,

$$\int_V |\nabla f|^2 \, dV = \int_V \nabla \cdot (f\nabla f) \, dV, \text{ since } \nabla^2 f = 0$$

$$= -\sum_i \int_{S_i} f\nabla f \cdot d\mathbf{S} - \int_{S_R} f\nabla f \cdot d\mathbf{S}.$$

To obtain the last line we have used the divergence theorem, with $d\mathbf{S}$ in the direction of the outward normal to the conductors.

For finite systems, any solution Φ is $\sim 1/r$ for large r. Hence $f \sim 1/R$ and $|\nabla f| \sim 1/R^2$ at points on S_R, and the integral over $S_R \to 0$ as $R \to \infty$. We then have

$$\int_V |\nabla f|^2 \, dV = -\sum_i \int_{S_i} f\nabla f \cdot d\mathbf{S}.$$

If the potential of a conductor is specified, then $f = \Phi_1 - \Phi_2 = 0$ on that conductor, and the surface integral over it is zero.

169

If the charge Q on a conductor is specified

$$Q = \int_S \sigma \, dS = \int_S \varepsilon_0 \, \mathbf{E} \cdot d\mathbf{S}, \text{ by (7.1)}$$

$$= -\varepsilon_0 \int_S \nabla \Phi \cdot d\mathbf{S}.$$

Since S is always an equipotential surface, $f = \Phi_1 - \Phi_2$ is constant over S, and

$$\int_S f \nabla f \cdot d\mathbf{S} = f_s \int_S \nabla f \cdot d\mathbf{S}$$

$$= f_s \left(\int_S \nabla \Phi_1 \cdot d\mathbf{S} - \int_S \nabla \Phi_2 \cdot d\mathbf{S} \right)$$

$$= f_s(-Q/\varepsilon_0 + Q/\varepsilon_0) = 0.$$

Thus if either the potential, or the charge, is specified for each conductor,

$$\int_V |\nabla f|^2 \, dV = 0,$$

and $\nabla f = 0$ everywhere in V. It follows that Φ_1 and Φ_2 can at most differ by a constant. Since $\Phi(r) \to 0$ as $r \to \infty$ this constant is zero: the two solutions must be identical.

Appendix C

Fields at the interface between materials

In this appendix we extend the arguments used in §7.1 for the conditions on the electrostatic field at the surface of a conductor, to obtain general matching conditions on time-dependent electric and magnetic fields at the interface between two materials.

The Maxwell equations for the macroscopic fields are

$$\mathbf{\nabla} \cdot \mathbf{D} = \rho_{\text{free}} \quad \text{(a),} \qquad \mathbf{\nabla} \times \mathbf{H} - \frac{\partial \mathbf{D}}{\partial t} = \mathbf{J}_{\text{free}} \quad \text{(b),}$$

$$\mathbf{\nabla} \cdot \mathbf{B} = 0 \quad \text{(c)} \qquad \mathbf{\nabla} \times \mathbf{E} + \frac{\partial \mathbf{B}}{\partial t} = 0 \qquad \text{(d),}$$

(C.1)

where ρ_{free} and $\mathbf{J}_{\text{free}}$ are the charge and current densities that have not been absorbed into the definitions of $\mathbf{D}$ and $\mathbf{H}$.

We consider a flat cylinder, a few atomic lengths in depth, enclosing a small macroscopic area δS of the interface. This may carry surface charge $\sigma \delta S$. We take the unit normal $\hat{\mathbf{n}}$ to be in the direction from material 1 to material 2 (Fig. C.1).

The integral form of (a) above,

$$\int \mathbf{D} \cdot \mathbf{dS} = \int \rho_{\text{free}} \, \mathrm{d}V,$$

gives

▶ $$(\mathbf{D}_2 \cdot \hat{\mathbf{n}}) - (\mathbf{D}_1 \cdot \hat{\mathbf{n}}) = \sigma,$$ (C.2)

since the flux through the sides of the cylinder is negligible on the macroscopic scale.

In particular, if $\sigma = 0$, the normal component of $\mathbf{D}$ is continuous across the interface.

Similarly, the integral form of (c),

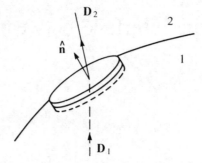

Fig. C.1 The flat cylinder, a few atomic spacings in depth, encloses a small (but macroscopic) element of the interface, $\delta\mathbf{S} = \hat{\mathbf{n}}\delta S$.

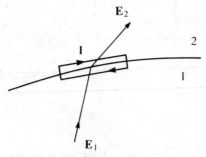

Fig. C.2 A path of integration at the interface between two materials.

$$\int \mathbf{B}\cdot d\mathbf{S} = 0,$$

gives

▶ $(\mathbf{B_2}\cdot\hat{\mathbf{n}}) - (\mathbf{B_1}\cdot\hat{\mathbf{n}}) = 0.$ (C.3)

Thus the normal component of **B** is *always* continuous across the interface.

The integral form of (d),

$$\oint \mathbf{E}\cdot d\mathbf{l} = -\int (\partial\mathbf{B}/\partial t)\cdot d\mathbf{S},$$

applied to the path shown in Fig. C.2, in which **l** is a small macroscopic length, and the segments through the surface are a few atomic lengths, gives

▶ $\mathbf{E_2}\cdot\mathbf{l} - \mathbf{E_1}\cdot\mathbf{l} = 0.$ (C.4)

The other contributions to the line integral, and the integral over the area of the loop, both include as a factor the atomic length through the surface, and are negligible. Since l can have any orientation in the surface, it follows that the component of **E** parallel to the interface is continuous. Similarly, using the integral form of (b),

$$\oint \mathbf{H} \cdot d\mathbf{l} = \int (\partial \mathbf{D}/\partial t + \mathbf{J}_{free}) \cdot d\mathbf{S},$$

it follows that the component of **H** parallel to the interface is continuous, unless there is a surface current density $\mathbf{j}_{surface}$ at the interface. In this case

$$\mathbf{H}_2 \cdot \mathbf{l} - \mathbf{H}_1 \cdot \mathbf{l} = l \mathbf{j}_{surface} \cdot \hat{\mathbf{t}}, \qquad (C.5)$$

where $\hat{\mathbf{t}} = \hat{\mathbf{n}} \times \hat{\mathbf{l}}$ is the unit normal to the loop (and a tangent to the surface). The condition (C.5) may be written

▶ $$\hat{\mathbf{n}} \times \mathbf{H}_2 - \hat{\mathbf{n}} \times \mathbf{H}_1 = \mathbf{j}_{surface} \qquad (C.6)$$

(using $\hat{\mathbf{l}} = \hat{\mathbf{t}} \times \hat{\mathbf{n}}$).

With the help of these matching conditions we can obtain overall solutions of Maxwell's equations, from solutions for (usually) homogeneous regions 1 and 2 on each side of the interface.

It is sometimes useful to neglect resistive losses in a first approximation, and assume a conductor is 'perfect', so that the electric field in its interior is always zero. From the Maxwell equation (d), there is then no time-dependent **B** field inside the conductor. It follows from (C.3) and (C.4) that just outside a perfect conductor, the normal component of the time-dependent **B** field and the tangential component of the **E** field must be zero. The time-dependent surface current density and the surface charge density will be given by (C.6) and (C.2) respectively. A 'surface current' is of course an idealisation, and corresponds to a very small skin depth.

Appendix D

Gaussian c.g.s. units

Gaussian c.g.s. units

SI units have been used in this book. However, this is not the only system of units possible. The units of mass, length, and time may be chosen differently, as may be the parameters ε_0 and μ_0, subject only to the condition that $\varepsilon_0 \mu_0 = 1/c^2$ where c is the velocity of light. Many texts and papers use *Gaussian c.g.s. units*, based on the centimetre, gramme and second. The unit of force in these units is called the *dyne* ($= 10^{-5}$ N) and the unit of energy is the *erg* ($= 10^{-7}$ J). The value of $4\pi\varepsilon_0$ is set $= 1$, and usually regarded as dimensionless so that charge is not an independent dimension: Coulomb's law becomes $\mathbf{F} = q_1 q_2 \hat{\mathbf{r}}/r^2$.

We then have $\mu_0 = 4\pi/c^2$. Also the magnetic field is made to have the same dimensions as the electric field by introducing a factor of c. Using lower case symbols for the fields expressed in Gaussian units, $\mathbf{E} \to \mathbf{e}$, $\mathbf{B} \to \mathbf{b}/c$, $\mathbf{J} \to \mathbf{j}$, $\rho \to \rho_{\mathrm{cgs}}$, and the Maxwell equations become more symmetrical in appearance:

$$\nabla \cdot \mathbf{e} = 4\pi\rho_{\mathrm{cgs}}, \quad \nabla \times \mathbf{b} - \frac{1}{c}\frac{\partial \mathbf{e}}{\partial t} = \frac{4\pi}{c}\mathbf{j}, \tag{D.1}$$

$$\nabla \cdot \mathbf{b} = 0, \quad \nabla \times \mathbf{e} + \frac{1}{c}\frac{\partial \mathbf{b}}{\partial t} = 0.$$

The Lorentz force equation is

$$\mathbf{F} = q[\mathbf{e} + (\mathbf{u}/c) \times \mathbf{b}]. \tag{D.2}$$

Turning now to the derived fields, $\mathbf{P} \to \mathbf{p}$ and $\mathbf{D} \to \mathbf{d}/4\pi$ (introducing a factor 4π), so that

$$\mathbf{D} = \varepsilon_0 \mathbf{E} + \mathbf{P} \to \mathbf{d} = \mathbf{e} + 4\pi\mathbf{p},$$

$$\nabla \cdot \mathbf{D} = \rho \to \nabla \cdot \mathbf{d} = 4\pi\rho_{\mathrm{cgs}}.$$

For a linear dielectric, $\mathbf{P} = \varepsilon_0 \chi_e \mathbf{E} \to \mathbf{p} = \chi_e^{cgs} \mathbf{e}$, so that $\chi_e^{cgs} = \chi_e/4\pi$.

Also $\mathbf{D} = \varepsilon_0 \varepsilon_r \mathbf{E} \to \mathbf{d} = \varepsilon_r \mathbf{e}$. In free space $\varepsilon_r = 1$ and $\mathbf{d} = \mathbf{e}$.

In the case of the magnetic derived fields, factors of c are introduced: $\mathbf{M} \to c\mathbf{m}$ and $\mathbf{H} \to c\mathbf{h}/4\pi$, so that

$$\mathbf{H} = \mathbf{B}/\mu_0 - \mathbf{M} \to \mathbf{h} = \mathbf{b} - 4\pi\mathbf{m}.$$

For a linear magnetic material, $\mathbf{M} = \chi_m \mathbf{H} \to \mathbf{m} = (\chi_m/4\pi)\mathbf{h}$, so that $\chi_m^{cgs} = \chi_m/4\pi$.

Also $\mathbf{B} = \mu_0 \mu_r \mathbf{H} \to \mathbf{b} = \mu_r \mathbf{h}$. In free space $\mu_r = 1$ and $\mathbf{b} = \mathbf{h}$.

The Maxwell equations for the macroscopic fields become

$$\nabla \cdot \mathbf{d} = 4\pi\rho_{cgs}, \quad \nabla \times \mathbf{h} - \frac{1}{c}\frac{\partial \mathbf{d}}{\partial t} = \frac{4\pi}{c}\mathbf{j}, \tag{D.3}$$

$$\nabla \cdot \mathbf{b} = 0, \quad \nabla \times \mathbf{e} + \frac{1}{c}\frac{\partial \mathbf{b}}{\partial t} = 0.$$

Units in the SI system and units in the Gaussian c.g.s. system are related by various powers of 10 (stemming from $1\ \text{m} = 10^2\ \text{cm}$, $1\ \text{kg} = 10^3\ \text{g}$) and factors of c. In particular, taking $c \approx 3 \times 10^8\ \text{m s}^{-1}$,

$1\ \text{C} \approx 3 \times 10^9$ Gaussian c.g.s. units of charge,

$1\ \text{T} = 10^4$ gauss exactly, where the *gauss* is the c.g.s. unit of magnetic field,

$1\ \text{V} \approx (1/300)$ Gaussian c.g.s. units of potential.

Further reading

Recommended as supplementary reading for its deep physical insight is:
Feynman, R. P. (1964), *The Feynman Lectures on Physics*, Volume II,
Reading, Mass: Addison-Wesley.
A somewhat more detailed treatment of wave guides and antennas
(together with excellent diagrams) will be found in:
Lorrain, P., Corson, D. R. and Lorrain, F. (1988), *Electromagnetic Fields
and Waves*, 3rd edition, San Francisco: Freeman.
An excellent graduate level text which includes a comprehensive discussion
of relativistic phenomena is:
Jackson, J. D. (1975), *Classical Electrodynamics*, 2nd edition, New York:
Wiley.
Two relevant volumes of 'Landau and Lifshitz' will appeal to readers with
strong theoretical interests:
Landau, L. D. and Lifshitz, E. M. (1975), *The Classical Theory of Fields*,
4th edition, Oxford: Pergamon.
Landau, L. D., Lifshitz, E. M. and Pitaevskiĭ (1984), *Electrodynamics of
Continuous Media*, 2nd edition, Oxford: Pergamon.
The microscopic description of the electric and magnetic properties of
solids is clearly presented in:
Ashcroft, N. W. and Mermin, N. D. (1976), *Solid State Physics*, New
York: Holt, Rinehart and Winston.

Answers to problems

Chapter 1

1.1 $r = (x^2 + y^2 + z^2)^{\frac{1}{2}}$; if $\mathbf{r} \neq 0$ then at least one of x or y or z must be $\neq 0$, and at least one factor of the product $\delta(\mathbf{r}) = \delta(x)\delta(y)\delta(z)$ is zero. Also, taking $dV = dx\,dy\,dz$, $\int \delta(\mathbf{r})\,dV = \int \delta(x)\,dx \int \delta(y)\,dy \int \delta(z)\,dz = 1$.

      For $x \neq 0$, $\delta_a(x) \to 0$ as $a \to 0$, and $\int_{-\infty}^{\infty} \delta_a(x)\,dx = 1$.

1.2 $\int \rho(\mathbf{r})\,dV = \int dV \int \rho_{a_1}(\mathbf{r} - \mathbf{r}')F(r')\,dV'$. Put $\mathbf{r}_1 = \mathbf{r} - \mathbf{r}'$, $\mathbf{r}_2 = \mathbf{r}'$. Then $dV\,dV' = dV_1\,dV_2$, and equation (1.9) gives $\int F(r_2)\,dV_2 = 1$. Hence the result.

1.5 0.94 mm s^{-1}.

1.6 6×10^7 m s^{-1}, 1.3×10^4 mm^{-3}, 8×10^{-9} s.

1.7   The atomic mass of Al is 26.98. 1 amu $= 1.66 \times 10^{-27}$ kg. Let N be the number of Al atoms collected per day from the negative electrode. 10^3 kg $= N \times 26.98 \times (1.66 \times 10^{-27}$ kg$)$.

      Charge transported by Al ions $= 3Ne = 1.07 \times 10^{10}$ C. 1 day $= 8.64 \times 10^4$ s. Hence the current due to the flow of Al^{3+} ions is 1.24×10^5 A. The same current is carried by the O^{2-} ions to maintain electrical neutrality, giving a total current of 2.48×10^5 A.

Chapter 2

2.1 $\mathbf{E} = -\nabla\Phi = \nabla(\mathbf{E}_0 \cdot \mathbf{r}) = \nabla(E_{0x}x + E_{0y}y + E_{0z}z) = (E_{0x}, E_{0y}, E_{0z})$.

2.2   The field is spherically symmetrical. Hence the flux through a spherical surface of radius r (with centre at the centre of the shell) is by Gauss's theorem

$$\int \mathbf{E} \cdot d\mathbf{S} = 4\pi r^2 E(r) = \begin{cases} 0 & \text{for } r < R, \\ Q/\varepsilon_0 & \text{for } r > R. \end{cases}$$

Hence the result.

2.3   As Problem 2.2, but here

$$4\pi r^2 E(r) = \begin{cases} (Q/\varepsilon_0)(r^3/a^3) & \text{for } r \leqslant R, \\ Q/\varepsilon_0 & \text{for } r > R. \end{cases}$$

Hence

$$E(r) = -\frac{d\Phi}{dr} = \begin{cases} (Q/4\pi\varepsilon_0)(r/a^3) & \text{for } r \leqslant R, \\ Q/4\pi\varepsilon_0 r^2 & \text{for } r > R. \end{cases}$$

Integrating,

$$\Phi(r) = \begin{cases} (Q/4\pi\varepsilon_0 a^3)(\tfrac{3}{2}a^2 - \tfrac{1}{2}r^2) & \text{for } r \leqslant R, \\ Q/4\pi\varepsilon_0 r & \text{for } r > R. \end{cases}$$

The constants of integration are determined by taking $\Phi(r) \to 0$ as $r \to \infty$, and $\Phi(r)$ continuous at $r = R$, since any discontinuity would imply an infinite electric field.

2.4   Since $\Phi_{\text{el}}(r)$ is spherically symmetrical, Poisson's equation reduces to

$$\frac{1}{r}\frac{d^2}{dr^2}(r\Phi_{\text{el}}) = \frac{e}{\pi\varepsilon_0 a_0^3}e^{-2r/a_0}.$$

This can be integrated in two steps to yield

$$\Phi_{\text{el}}(r) = \frac{e}{4\pi\varepsilon_0}e^{-2r/a_0}\left(\frac{1}{r} + \frac{1}{a_0}\right) + A + \frac{B}{r},$$

where A and B are integration constants. Setting $\Phi_{\text{el}}(r) \to 0$ as $r \to \infty$ implies $A = 0$; and $B = -e/4\pi\varepsilon_0$, appropriate to the total electronic charge. Adding the potential due to the nucleus cancels this last term, to give the atomic potential.

2.5   Away from the ends of the wire, the field has axial symmetry, and the flux per unit length through a cylindrical surface of radius ρ, centred on the wire, is $2\pi\rho E(\rho)$, which by Gauss's theorem equals $Q/l\varepsilon_0$. Hence $E(\rho) = -\partial\Phi/\partial\rho = (Q/l)/2\pi\varepsilon_0\rho$. Integrating, $\Phi(\rho) = -(Q/2\pi\varepsilon_0 l)\ln\rho +$ constant. These expressions are only valid for $\rho \ll l$.

2.6   The transverse velocity v_{T} of the electron in a field $-E$ is given by $m_e\,dv_{\text{T}}/dt = eE$. After a time t in the field, $v_{\text{T}} = eEt/m$. Approximately, $t = d/v_{\text{L}}$, where $d = 4$ cm and v_{L} is the longitudinal (beam) velocity. Then the angle of deflection is given by $\tan\theta = v_{\text{T}}/v_{\text{L}} = eEd/m_e v_{\text{L}}^2$. $\theta = 10°$, and $m_e v_{\text{L}}^2/2$ is the beam energy 10 keV, giving $E = 88$ kV m^{-1}.

2.7   Using the method of Problem 2.5, the radial electric field at the fringe of the beam is $E_\rho = -(\pi\rho^2 ne)/2\pi\varepsilon_0\rho$, where ρ is the radius of the beam, and n the electron density, giving $E_\rho = -59$ V m^{-1} (see Problem 1.6 and answer). In the transit time t, $\Delta\rho \approx \tfrac{1}{2}(eE_\rho/m_e)\,t^2 = 0.33$ mm.

2.8   Mean value $= \int_0^R \Phi(r)\,4\pi r^2\,dr/\tfrac{4}{3}\pi R^3$. Taking $\Phi(r)$ from Problem 2.3 gives the stated result. The mean electrostatic energy of the nth added proton in the averaged field of $(n-1)$ protons is $\tfrac{6}{5}(n-1)e^2/4\pi\varepsilon_0 R$. The total electrostatic energy is

$$[1+2+\ldots+(Z-1)]\tfrac{6}{5}e^2/4\pi\varepsilon_0 R = Z(Z-1)\tfrac{3}{5}e^2/4\pi\varepsilon_0 R.$$

1.06×10^3 MeV.

2.9   Field energy $= \dfrac{\varepsilon_0}{2}\displaystyle\int_a^\infty E^2 4\pi r^2\,dr = \dfrac{\varepsilon_0}{2}\displaystyle\int_a^\infty \left(\dfrac{e}{4\pi\varepsilon_0 r}\right)^2 4\pi r^2\,dr = e^2/8\pi\varepsilon_0 a.$

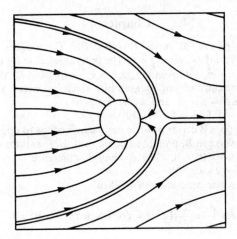

Answer 2.11

Setting $e^2/8\pi\varepsilon_0 a = m_e c^2$ gives $a = 1.4$ fm.

2.11   See figure. The proportion of the area containing field lines which end on the cylinder is $\sim 44\%$.

2.12   From Problem 2.5, at a distance ρ from the wire the potential $\Phi(\rho) = A\ln\rho + B$. The constants A and B are determined since $\Phi = 3.5$ kV at $\rho = 1$ μm and $\Phi = 0$ at $\rho = 1$ cm. Then $E(\rho) = -\partial\Phi/\partial\rho = (507/\rho)$ V. For an electron to gain 4 eV in travelling 10^{-7} m, the field $E > 4 \times 10^7$ V m^{-1}, and this occurs for $\rho < 12.7$ μm. In reaching the wire there would be ~ 27 doublings giving $2^{27} = 1.3 \times 10^8$ electrons.

Chapter 3

3.1   If the origin is shifted to $\mathbf{a}$, and $\mathbf{r}' = \mathbf{r} - \mathbf{a}$,

$$\mathbf{p}' = \int \rho'(\mathbf{r}')\,\mathbf{r}'\,dV' = \int \rho(\mathbf{r})(\mathbf{r}-\mathbf{a})\,dV = \mathbf{p} - \mathbf{a}\int \rho(\mathbf{r})\,dV.$$

Hence if $Q = \int \rho(\mathbf{r})\,dV = 0$, $\mathbf{p}' = \mathbf{p}$.

3.3   Use equations (3.5) and (3.10).

3.4   The field of the positive ion is $\mathbf{E} = (e/4\pi\varepsilon_0 r^2)\hat{\mathbf{r}}$. The configuration of minimum energy has the dipole moment aligned with the field, and the energy is $-\mathbf{p}\cdot\mathbf{E} = -pE$. In this orientation, the force $\mathbf{F} = (\mathbf{p}\cdot\nabla)\mathbf{E} = p\,\partial E/\partial r$ is attractive. To reverse the orientation requires energy $2pE = 2.8 \times 10^{-19}$ J $= 1.8$ eV. At room temperature, $k_B T \approx (1/40)$ eV.

3.5   Write

$$\rho(\mathbf{r}) = q[\delta(r_1 - a)\,\delta(r_2)\,\delta(r_3) + \delta(r_1 + a)\,\delta(r_2)\,\delta(r_3)$$
$$- \delta(r_1)\,\delta(r_2 - a)\,\delta(r_3) - \delta(r_1)\,\delta(r_2 + a)\,\delta(r_3)]$$

in equations (3.2) and (3.7).

3.6 $\mathbf{p} = \int \mathbf{r}\rho(\mathbf{r})\,dV = \int dV \mathbf{r} \int \rho_{at}(\mathbf{r} - \mathbf{r}')F(\mathbf{r}')\,dV'$. Put $\mathbf{r}_1 = \mathbf{r} - \mathbf{r}'$, $\mathbf{r}_2 = \mathbf{r}'$. Then $dV\,dV' = dV_1\,dV_2$ and $\mathbf{p} = \iint (\mathbf{r}_1 + \mathbf{r}_2)\rho_{at}(\mathbf{r}_1)F(\mathbf{r}_2)\,dV_1\,dV_2 = \int \mathbf{r}_1\rho_{at}(\mathbf{r}_1)\,dV_1$, since $\int F(\mathbf{r}_2)\,dV_2 = 1$, and (since we take the smoothing function F to be spherically symmetric) $\int \mathbf{r}_2 F(\mathbf{r}_2)\,dV_2 = 0$.

Chapter 4

4.1 $m\,dv/dt = (-e)\,\mathbf{v} \times \mathbf{B}$ and $\mathbf{B} = (0, 0, B)$. Component equations are
$m\ddot{x} = -eB\dot{y}, m\ddot{y} = eB\dot{x}, m\ddot{z} = 0$. The general solution for motion in a plane
$z = z_0$ is $x = x_0 + a\cos\omega(t - t_0)$, $y = y_0 + a\sin\omega(t - t_0)$ (where x_0, y_0, z_0,
t_0 and a are constants of integration), which is a circle. More generally, if
$\dot{z} = u$ then $z = z_0 + ut$, and we have a helix.
        For $B = 1$ T, $\omega = 0.176 \times 10^{12}$ s^{-1}.

4.3   The force $Q\mathbf{v} \times \mathbf{B}$ drives positive and negative ions to opposite sides of the
pipe transverse to $\mathbf{B}$, until the electric field they produce just compensates
and $Q(\mathbf{E} + \mathbf{v} \times \mathbf{B}) = 0$. The potential difference across the pipe is
$Ed = vBd = 2 \times 10^{-2}$ V.

4.4   Take the scalar product with $\mathbf{v}$. Then

$$m\mathbf{v} \cdot \frac{d\mathbf{v}}{dt} = \frac{d}{dt}(\tfrac{1}{2}mv^2) = Q\mathbf{v} \cdot \mathbf{E} \text{ since } \mathbf{v} \cdot (\mathbf{v} \times \mathbf{B}) = 0.$$

Integrating,

$$\tfrac{1}{2}mv_1^2 - \tfrac{1}{2}mv_0^2 = \int_{t_0}^{t_1} Q\mathbf{v} \cdot \mathbf{E}\,dt = Q\int_{\mathbf{r}_0}^{\mathbf{r}_1} \mathbf{E} \cdot d\mathbf{r} = Q(\Phi(\mathbf{r}_0) - \Phi(\mathbf{r}_1))$$

or
$$\tfrac{1}{2}mv_1^2 + Q\Phi(\mathbf{r}_1) = \tfrac{1}{2}mv_0^2 + Q\Phi(\mathbf{r}_0) = \text{constant}.$$

4.5 1.5×10^{-3} T.
        Note that this field is technically more convenient than the large electric
field of Problem 2.6.

4.6   The flux through a spherical surface of radius r is $\int \mathbf{B} \cdot d\mathbf{S} = 4\pi r^2 B(r)$. By
Gauss's theorem this is zero. Hence $B(r) = 0$ for all r.

4.7   The cable has cylindrical symmetry. There can be no radial component to
$\mathbf{B}$ (cf. Problem 4.6), and no component along the axis of the cylinder.
(Consider the current flow as a superposition of filaments.) Hence the field
lines are circles around the axis, and for a circle of radius ρ Ampère's law
gives:

$$\oint \mathbf{B} \cdot d\mathbf{l} = 2\pi\rho B_\phi(\rho) = \mu_0 \begin{cases} I(\rho^2/a^2), & 0 < \rho < a \\[2mm] I, & a < \rho < b_1 \\[2mm] \left[I - I\left(\dfrac{\rho^2 - b_1^2}{b_1^2 - b_2^2}\right)\right], & b_1 < \rho < b_2 \\[2mm] 0, & \rho > b_2. \end{cases}$$

4.8   The current distribution is obtained from $\nabla \times \mathbf{B} = \mu_0 \mathbf{J}$. Here $\mathbf{B} = (B(z), 0, 0)$
so that $\nabla \times \mathbf{B} = (0, \partial B_z/\partial z, 0)$.
        $\partial B_z/\partial z = 0$ except at $z = 0$ and $z = a$, where it is singular. We can
integrate J_y across these singularities, to give surface current densities
$\pm B/\mu_0$ in the planes $z = 0$, $z = a$.

4.9   A small displacement $y(x)$ at a point x on the wire is given by $T\,d^2y/dx^2$
$= -IB$, which with $y(0) = y(l) = 0$ has solution $y(x) = (IB/2T)x(l - x)$.
The extension is given approximately by $\delta l = \tfrac{1}{2}\int_0^l (dy/dx)^2\,dx$ and $\delta l/l =$
$T/\pi a^2\lambda$. The results follow and $T = 23.6$ N, $d = y(l/2) = 5$ mm.

Chapter 5

5.1 Consider a small area dS in the plane of the loop. At time t the flux through it is $\mathbf{B} \cdot d\mathbf{S} = B \, dS \cos \omega t$, and the total flux through the loop is
$\mathscr{F} = \int B \, dS \cos \omega t = BS \cos \omega t$.
Hence $\mathscr{E} = -d\mathscr{F}/dt = BS\omega \sin \omega t$.

5.2 When the loop carries a current I each element $d\mathbf{l}$ is subject to a magnetic force $I \, d\mathbf{l} \times \mathbf{B}$, implying a net torque $\mathbf{M} = I \oint \mathbf{r} \times (d\mathbf{l} \times \mathbf{B})$ on the loop. In isolation the torque would slow the rotation. To maintain constant rotation an external torque $-\mathbf{M}$ must be applied, and this does work at a rate $-\mathbf{M} \cdot \boldsymbol{\omega} = -I \oint (\mathbf{r} \cdot \mathbf{B}) \, \boldsymbol{\omega} \cdot d\mathbf{l} + I \mathbf{B} \cdot \boldsymbol{\omega} \oint \mathbf{r} \cdot d\mathbf{l}$, expanding the triple vector product. Now $\oint \mathbf{r} \cdot d\mathbf{l} = 0$, so that $-\mathbf{M} \cdot \boldsymbol{\omega} = -I \oint (\mathbf{r} \cdot \mathbf{B}) \, \boldsymbol{\omega} \cdot d\mathbf{l} = I \int \boldsymbol{\omega} \times \mathbf{B} \cdot d\mathbf{S}$ (using Stokes's theorem) $= I\omega BS \sin \omega t = I\mathscr{E} = I^2 R$. (The expression for $\mathbf{M}$ simplifies if the loop is a rectangle.)

5.3 From symmetry, and since there are no sources of electric charge, the electric field lines are circles. For such a circle of radius ρ, $\oint \mathbf{E} \cdot d\mathbf{l} = 2\pi\rho E(\rho, t)$ and the result follows from equation (5.4).

The Lorentz equation is $d\mathbf{p}/dt = (-e)(\mathbf{E} + \mathbf{v} \times \mathbf{B})$. For motion in a circle, the tangential component is $dp_\phi/dt = (-e) E$, giving

$$p_\phi = (-e) \int_0^t E \, dt = (e/2\pi\rho) \mathscr{F}(\rho, t).$$

For motion in a circle, the radial component inwards is $-dp_\rho/dt = p_\phi \, d\phi/dt = p_\phi(v/\rho)$. This is equal to the inward magnetic force Bev if $B = p_\phi/e\rho = \mathscr{F}(\rho, t)/2\pi\rho^2$, and $\mathscr{F}(\rho, t)/\pi\rho^2 = \int B \, dS/\pi\rho^2 = $ average field.

48 MeV/c. The electron momentum is relativistic. However, the Lorentz equation remains valid (Ch. 19).

5.4 The magnetic field must also have spherical symmetry and is therefore zero (Problem 4.6). The charge $Q(r, t)$ within a sphere of radius r gives a time-dependent field $E(r, t) = Q(r, t)/4\pi\varepsilon_0 r^2$. Equation (5.7b) becomes $4\pi r^2 J(r) = \partial Q(r, t)/\partial t$, and expresses charge conservation.

Chapter 6

6.1 In a plane wave $|\mathbf{E}| = c|\mathbf{B}|$, $|\mathbf{N}| = c\varepsilon_0 |\mathbf{E}|^2$. Energy per second $= c\varepsilon_0 |\mathbf{E}|^2 \times$ area of beam $= 10^{-3}$ W. Hence $\overline{|\mathbf{E}^2|} = (690 \text{ V m}^{-1})^2$. For linearly polarised light $|\mathbf{E}|_{max} = \sqrt{2} \times (690 \text{ V m}^{-1})$; for circularly polarised light $|\mathbf{E}|_{max} = 690 \text{ V m}^{-1}$.

6.2 $$f(z - ct) = \frac{a}{\sqrt{\pi}} \int_{-\infty}^{\infty} e^{-a^2 p^2} \cos p(z - ct) \cos k_0(z - ct) \, dp.$$

Use $\cos A \cos B = \frac{1}{2}[\cos(A + B) + \cos(A - B)]$. The two terms contribute equally, and after changes of integration variables give

$$f(z - ct) = \frac{a}{\sqrt{\pi}} \int_{-\infty}^{\infty} e^{-a^2(k - k_0)^2} \cos k(z - ct) \, dk.$$

Because of the properties of the exponential function, wave vectors in the range $|k - k_0| \lesssim 1/a$ give the main contribution, i.e., frequencies in the band $c(k_0 - 1/a) < \omega < c(k_0 + 1/a)$.

6.3   Neglecting the magnetic force, $m\ddot{\mathbf{r}} = (-eE_0 \cos(kz - \omega t), 0, 0)$. In this
        approximation there is no motion in the y- or z-directions, and $m\ddot{x} = -eE_0 \cos \omega t$. Hence $\dot{x} = -(eE_0/m\omega)\sin \omega t$, harmonic motion with
        maximum velocity $eE_0/m\omega = eE_0 \lambda/2\pi mc = 0.064$ m s^{-1}.
            The ratio of magnetic to electric force is $|\mathbf{v} \times \mathbf{B}|/|\mathbf{E}|$, and since $|\mathbf{E}| = c|\mathbf{B}|$
        this equals $\dot{x}/c < 10^{-9}$.

6.4   Equations (6.17) and (6.18) are easily modified to the new geometry, and

$$\frac{\omega^2}{c^2} = \pi^2 \left(\frac{l^2}{L_x^2} + \frac{m^2}{L_y^2} + \frac{n^2}{L_z^2}\right).$$

If $L_x = 0.2$ m, $L_y = 0.18$ m, $L_z = 0.15$ m, the lowest frequency has $l = m = 1$, $n = 0$, which gives $\omega = 7 \times 10^9$ s^{-1}, $v = \omega/2\pi = 1.12 \times 10^9$ Hz.

$$\mathbf{E} = \left(0, 0, \alpha \sin\left(\frac{\pi x}{L_x}\right)\sin\left(\frac{\pi y}{L_y}\right)\cos \omega t\right),$$

$$\mathbf{B} = \left(-\frac{\pi\alpha}{L_y \omega}\sin\left(\frac{\pi x}{L_x}\right)\cos\left(\frac{\pi y}{L_y}\right)\sin \omega t, \frac{\pi\alpha}{L_x \omega}\cos\left(\frac{\pi x}{L_x}\right)\sin\left(\frac{\pi y}{L_y}\right)\sin \omega t, 0\right).$$

6.5   Note

$$\int_0^L \sin^2\left(\frac{l\pi x}{L}\right)dx = \int_0^L \cos^2\left(\frac{l\pi x}{L}\right)dx = \frac{L}{2}.$$

From equation (6.15), the electric field energy of the waveform (6.17) is
$\mathscr{E}_T \cos^2 \omega t$, with $\mathscr{E}_T = (\mathscr{E}_0 L^3/16)(\alpha_x^2 + \alpha_y^2 + \alpha_z^2)$.
    The normal component of the Poynting vector is zero on the sides of
the cube, since the electric field is normal to the surface. Hence the total
energy is constant. The magnetic energy is therefore $\mathscr{E}_T \sin^2 \omega t$, and the
total energy is $\mathscr{E}_T$.

Chapter 7

7.1   For $a < \rho < b$, $E(\rho) = (Q/l)/2\pi\varepsilon_0 \rho$ (cf. Problem 2.5), and

$$\Phi = -(Q/2\pi\varepsilon_0 l)\ln(\rho/b),$$

setting $\Phi(b) = 0$.
        $C = Q/\Phi(a) = 2\pi\varepsilon_0 l/\ln(b/a)$.

7.3   Using equation (7.1),

$$\sigma(\theta) = \varepsilon_0 E_r(\theta) = -\varepsilon_0 \partial\Phi/\partial r|_{r=a} = 3\varepsilon_0 E_0 \cos \theta.$$

The dipole moment of the induced surface charge is, by symmetry, along
the field direction, and is of magnitude

$$\int \sigma(\theta) a \cos \theta \, dS = 3\varepsilon_0 E_0 a^3 \int_0^\pi \cos^2 \theta 2\pi \sin \theta \, d\theta$$

$$= 4\pi\varepsilon_0 a^3 E_0.$$

7.4   By the superposition principle, the potential of the charged sphere $Q/4\pi\varepsilon_0 r$
        is to be added.

7.5   Assume a charge q per unit length on one wire (1) induces a charge $-q$ per unit length on the other (2). If $d \gg a$, the potential at one wire due to the other is very nearly constant over the wire, and

$$V_1 \approx -(q/2\pi\varepsilon_0)\ln a + (q/2\pi\varepsilon_0)\ln d$$
$$V_2 \approx (q/2\pi\varepsilon_0)\ln a - (q/2\pi\varepsilon_0)\ln d$$

(up to an irrelevant constant). The potential difference between the wires is

$$V = V_1 - V_2 \approx (q/\pi\varepsilon_0)\ln(d/a).$$

The capacitance per unit length is $q/V = \pi\varepsilon_0/\ln(d/a)$.

7.6   The electrostatics of the model is like a parallel plate capacitor, and the potential difference is $\sigma d/\varepsilon_0 = 9$ V.

7.7   If $\Phi = -E_0\rho\cos\phi = -E_0 x$, $\mathbf{E} = -\nabla\Phi = (E_0, 0, 0)$. The given potential for the wire in a uniform field satisfies Laplace's equation outside the wire, is constant over the surface of the wire, and behaves correctly at large (but not too large!) distances. The term $-(q/2\pi\varepsilon_0)\ln\rho$ gives the correct flux leaving the wire. The electric field at the surface of the wire is

$$E_\rho = -\partial\Phi/\partial\rho|_a = 2E_0\cos\phi + q/2\pi\varepsilon_0 a.$$

This is positive for all ϕ if $q > q_{\min} = 4\pi\varepsilon_0 aE_0$. Since field lines do not cross, field lines leaving the wire when $q = q_{\min}$ will form a band of width $q_{\min}/\varepsilon_0 E_0 = 4\pi a$ at large distances, by Gauss's theorem.

Chapter 8

8.1 $\tau = 3.2 \times 10^{-14}$ s, distance $= 3.4 \times 10^{-8}$ m $= 340$ Å.

8.2   Aluminium wire 3.2×10^{-2} Ω. Bismuth wire 1.4 Ω. The electric field strength in a wire of length l, and resistance R, carrying current I is $E = IR/l$. $E_{\text{Al}} = 3.2 \times 10^{-2}$ V m^{-1}, $E_{\text{Bi}} = 1.4$ V m^{-1}.

By Gauss's theorem, there is charge at the junction $= \varepsilon_0(E_{\text{Al}} - E_{\text{Bi}})S = -9.3 \times 10^{-18}$ C. This corresponds to about 58 electrons.

8.3   Consider one conductor (1) to be at potential $\frac{1}{2}V$, and the other (2) at potential $-\frac{1}{2}V$. At a point distance ρ_1 from (1), ρ_2 from (2), the potential is approximately

$$\Phi = \frac{V}{2\ln(d/a)}(-\ln\rho_1 + \ln\rho_2)$$

(cf. Problem 7.5).

The electric field at the surface of (1) is

$$E_\rho = -\partial\Phi/\partial\rho_1|_a \approx V/2a\ln(d/a),$$

and hence the current per unit length out of (1) (and into (2)) is $i = 2\pi a\sigma E_\rho = \sigma\pi V/\ln(d/a)$, and the resistance per unit length is $V/i = \ln(d/a)/\pi\sigma$.

8.4   At the surface of the wire the tangential component of $\mathbf{E}$ is continuous, and inside the wire is $E = J/\sigma = I/\pi a^2\sigma$. Outside the wire it is given, correctly, by $-\partial\Phi/\partial z|_{\rho=a}$. Also $\Phi(b, z) = 0$ for all z.

$$\sigma(\rho, z) = \varepsilon_0 E_n = -\varepsilon_0\partial\Phi/\partial\rho|_{\rho=a}.$$

Chapter 9

9.1   Suppose $A_z(\mathbf{r}) = 0$. Then $B_x = \partial A_z/\partial y - \partial A_y/\partial z = -\partial A_y/\partial z$ giving
$A_y(x, y, z) = -\int_{-\infty}^{z} B_x(x, y, z')\,\mathrm{d}z' + \text{constant}$.
       Similarly $A_x(x, y, z) = \int_{-\infty}^{z} B_y(x, y, z')\,\mathrm{d}z' + \text{constant}$. Then

$$\frac{\partial A_y}{\partial x} - \frac{\partial A_x}{\partial y} = -\int_{-\infty}^{z} \left(\frac{\partial B_x}{\partial x} + \frac{\partial B_y}{\partial y}\right)\mathrm{d}z'$$

$$= \int_{-\infty}^{z} \frac{\partial B_z}{\partial z'}\,\mathrm{d}z'\ (\text{since } \nabla \cdot \mathbf{B} = 0) = B_z.$$

Hence this vector potential gives the magnetic field.

9.2 $\mathbf{A} = B_0(0, x, 0)$, or $\mathbf{A} = \frac{1}{2}B_0(-y, x, 0)$, for example. The latter expression $= \frac{1}{2}\mathbf{B}_0 \times \mathbf{r}$.

9.3 $\chi(\mathbf{r})$ must satisfy $\nabla^2 \chi = -\alpha(\mathbf{r})$, which has the solution

$$\chi(\mathbf{r}) = \frac{1}{4\pi} \int \frac{\alpha(\mathbf{r}')}{|\mathbf{r} - \mathbf{r}'|}\,\mathrm{d}V'$$

(cf. equation (2.19)).

9.4   If the circle is parameterised as $(a\cos\phi, a\sin\phi, 0)$ then $\mathrm{d}\mathbf{l} = (-a\sin\phi, a\cos\phi, 0)\,\mathrm{d}\phi$. At $\mathbf{r} = (0, 0, z)$, the Biot–Savart law gives

$$\mathbf{B}(0, 0, z) = \frac{\mu_0 I}{4\pi} \int_0^{2\pi} \frac{(-a\sin\phi, a\cos\phi, 0) \times (-a\cos\phi, -a\sin\phi, z)}{(a^2 + z^2)^{\frac{3}{2}}}\,\mathrm{d}\phi$$

$$= \frac{\mu_0 I}{4\pi} \int_0^{2\pi} \frac{(az\cos\phi, az\sin\phi, a^2)}{(a^2 + z^2)^{\frac{3}{2}}}\,\mathrm{d}\phi$$

$$= (0, 0, \mu_0 I a^2/2(a^2 + z^2)^{\frac{3}{2}}).$$

Consider two such coils, at $\pm z_0$,

$$B(z) = \frac{\mu_0 I a^2}{2}\left[\frac{1}{[a^2 + (z_0 - z)^2]^{\frac{3}{2}}} + \frac{1}{[a^2 + (z_0 + z)^2]^{\frac{3}{2}}}\right].$$

Expand in powers of z using Taylor's theorem:

$$B(z) = \frac{\mu_0 I a^2}{(a^2 + z^2)^{\frac{3}{2}}}\left[1 + \frac{3(4z_0^2 - a^2)}{2(a^2 + z_0^2)^2}z^2 + \dots\right].$$

The coefficient of z^2 is zero if $2z_0 = a$.

9.5   At $(0, 0, z)$, the Biot–Savart law gives

$$B_z = \frac{\mu_0 I}{4\pi} \int_{-\infty}^{\infty} \frac{a^2\,\mathrm{d}\phi}{[z^2 + (\alpha\phi - z)^2]^{\frac{3}{2}}} = \frac{\mu_0 I}{4\pi} \int_{-\infty}^{\infty} \frac{\mathrm{d}q}{(1 + q^2)^{\frac{3}{2}}},$$

making the change of variable $q = (\alpha\phi - z)/a$.

9.6   (a) Use $\nabla \cdot (F\mathbf{J}) = \mathbf{J} \cdot \nabla F + F\nabla \cdot \mathbf{J} = \mathbf{J} \cdot \nabla F$ since $\nabla \cdot \mathbf{J} = 0$. Then $\int \nabla \cdot (F\mathbf{J})\,\mathrm{d}V = \int F\mathbf{J} \cdot \mathrm{d}\mathbf{S}$ by the divergence theorem, and the surface integral is zero since $\mathbf{J}$ is localised.

9.8   Torque $= \int \mathbf{r} \times (\mathbf{J} \times \mathbf{B}) \, dV = \int (\mathbf{r} \cdot \mathbf{B}) \mathbf{J} \, dV$,

$$\mathbf{m} \times \mathbf{B} = -\frac{1}{2} \int \mathbf{B} \times (\mathbf{r} \times \mathbf{J}) \, dV = \frac{1}{2} \int (\mathbf{r} \cdot \mathbf{B}) \mathbf{J} \, dV - \frac{1}{2} \int (\mathbf{B} \cdot \mathbf{J}) \mathbf{r} \, dV$$

$$= \int (\mathbf{r} \cdot \mathbf{B}) \mathbf{J} \, dV \text{ also.}$$

9.9   The nucleus is subject to a torque $\mathbf{m} \times \mathbf{B} = \gamma \mathbf{J} \times \mathbf{B}$, so that $d\mathbf{J}/dt = \gamma \mathbf{J} \times \mathbf{B}$.
       Take $\mathbf{B} = (0, 0, B)$. Then $\dot{J}_x = \gamma B J_y$, $\dot{J}_y = -\gamma B J_x$, $\dot{J}_z = 0$. These equations
       have the solution $J_x = J_1 \sin \omega t$, $J_y = J_1 \cos \omega t$, $J_z = J_2$, where $\omega = \gamma B$ and
       $J_1^2 + J_2^2 = J^2$.

9.10   Between the conductors, $B = \mu_0 I / 2\pi\rho$ (Problem 4.7). Hence $U = \int_a^{b_1} (B^2/2\mu_0) 4\pi\rho \, d\rho = (\mu_0 I^2/4\pi) \ln(b_1/a)$. If $a \ll b_1$ and $(b_2 - b_1) \ll b_1$, the
       magnetic energy in the conductors is negligible; the self inductance follows
       since $U = \frac{1}{2} L I^2$.

9.11 $U = \frac{1}{2} L_{11} I_1^2 + L_{12} I_1 I_2 + \frac{1}{2} L_{22} I_2^2$

$$= \frac{1}{2} L_{11} \left(I_1 + \frac{L_{12}}{L_{11}} I_2 \right)^2 + \frac{1}{2} \left(L_{22} - \frac{L_{12}^2}{L_{11}} \right) I_2^2.$$

       $(I_1 + L_{12} I_2 / L_{11})$ can be made zero. Hence $L_{22} - L_{12}^2 / L_{11}^2 \geqslant 0$ since $U \geqslant 0$,
       i.e., $L_{12} < \sqrt{(L_{11} L_{22})}$.

9.12   Parameterise the circles as $(R_1 \cos \theta_1, R_1 \sin \theta_1, 0)$ and $(R_2 \cos \theta_2, R_2 \sin \theta_2, 0)$
       and put $\theta = \theta_2 - \theta_1$; $d\theta_1 \, d\theta_2 \rightarrow d\theta_1 \, d\theta$.
       For large α,

$$\frac{1}{[2(1 - \cos \theta) + \alpha^2]^{\frac{1}{2}}} \approx \frac{1}{\alpha} - \frac{1}{2} \left[\frac{2(1 - \cos \theta)}{\alpha^3} \right].$$

       $G(0.01) = 4.69$, $G(0.5) = 0.89$, $G(1) = 0.39$, $G(3) = 0.04$.

9.13   When a current I flows in the outer coil, the field is $\mu_0 n I$ and hence the flux
       through the nz turns of the inner coil is $nz\pi a^2 (\mu_0 n I) = L_{12} I$.

9.15   When a current I flows, the field energy outside the wire per unit length is
       approximately $\int_a^{\rho_c} (B^2/2\mu_0) 2\pi\rho \, d\rho$, where $B \sim \mu_0 I / 2\pi\rho$ and the cut-off
       distance $\rho_c \sim b$. The result follows if the field energy inside the conductors
       is neglected, using $U = \frac{1}{2} L I^2$.

9.16 $U = \frac{1}{2} L_{11} I_1^2 + L_{12} I_1 I_2 + \frac{1}{2} L_{22} I_2^2$.
       In a displacement $\delta \mathbf{R}$

$$\delta U = L_{11} I_1 \delta I_1 + L_{12} (I_1 \delta I_2 + I_2 \delta I_1) + L_{22} I_2 \delta I_2 + I_1 I_2 \delta \mathbf{R} \cdot \nabla L_{12}$$

$$= I_1 (L_{11} \delta I_1 + L_{12} \delta I_2) + I_2 (L_{22} \delta I_2 + L_{21} \delta I_1) + I_1 I_2 \delta \mathbf{R} \cdot \nabla L_{12}.$$

       But since the flux through the loops does not change, $(L_{11} \delta I_1 + L_{12} \delta I_2) = 0$ and $(L_{22} \delta I_2 + L_{21} \delta I_1) = 0$. The result follows from conservation of
       energy.

Chapter 10

10.1   1 mole $(6.02 \times 10^{23}$ atoms$)$ of a monatomic gas at STP occupies
       22.4×10^{-3} m^3. Hence for $\alpha \sim (1 \text{ Å})^3 = 10^{-30}$ m^3, $\chi_e = 4\pi N\alpha \sim 10^{-4}$. For
       the solid $\chi_e \sim 4\pi N\alpha$ and $N \sim$ (atomic volume)$^{-1}$, so that $\chi_e \sim 1$.

10.2 $$\bar{p} = \int_0^\pi p \cos \theta f(\theta) \, d\theta$$

$$= \frac{\displaystyle\int_0^\pi p \cos \theta \exp\left(pE \cos \theta / k_B T\right) \sin \theta \, d\theta}{\displaystyle\int_0^\pi \exp\left(pE \cos \theta / k_B T\right) \sin \theta \, d\theta}$$

$$= p \int_{-1}^1 x \exp\left(pEx/k_B T\right) dx \bigg/ \int_{-1}^1 \exp\left(pEx/k_B T\right) dx.$$

The result follows, after integrating the numerator by parts.
For small z, $\coth z \approx 1/z + z/3$.
Then $P = N\bar{p} \approx Np^2 E/3k_B T$, and $\chi_e = P/\varepsilon_0 E$.

10.3 $\chi_e = 1.17 \times 10^{-4}$ (20 °C $\approx$ 293 K).

10.4   The boundary conditions on the tangential components of $\mathbf{E}$ and the normal components of $\mathbf{D}$ give $E_1 \sin \theta_1 = E_2 \sin \theta_2$, $\varepsilon_1 E_1 \cos \theta_1 = \varepsilon_2 E_2 \cos \theta_2$, respectively, so that $\tan \theta_2 = (\varepsilon_2/\varepsilon_1) \tan \theta_1$.

10.5   The equivalent polarisation charge is a surface charge of density $\mathbf{P} \cdot \mathbf{n} = P \cos \theta$ (taking $\mathbf{P}$ in the z-direction). Assume

$$\Phi = \begin{cases} Ar \cos \theta + \text{constant}, & r < a \\ B \cos \theta / r^2, & r > a. \end{cases}$$

At $r = a$, the boundary conditions on $\mathbf{E}$ are satisfied if

$$Aa = B/a^2 \text{ (tangential component)},$$

$$2B/a^3 + A = P/\varepsilon_0 \text{ (normal component)}.$$

Hence $A = P/3\varepsilon_0$, $B = Pa^3/3\varepsilon_0$.

10.6 2.5×10^5 electrons per second.
The capacitance of the spiral roll is approximately double.

10.7 $V_h/V_v = C_v/C_h$ since Q remains the same.

$$C_v = (A/2)\varepsilon_0 \varepsilon_r/d + (A/2)\varepsilon_0/d \text{ (in parallel)}$$
$$C_h^{-1} = (A\varepsilon_0 \varepsilon_r/\tfrac{1}{2}d)^{-1} + (A\varepsilon_0/\tfrac{1}{2}d)^{-1} \text{ (in series)}.$$

Hence the result.

10.8 $C = (L^2 - bL)\varepsilon_0/d + bL\varepsilon_0\varepsilon_r/d$ and $U = Q^2/2C$. By conservation of energy, force $F = -\partial U/\partial b = (Q^2/2C^2)\partial C/\partial b$. If V is held constant by a battery, $U = CV^2/2$, and in a displacement δb the battery supplies work $V\delta Q$. The energy equation becomes $\delta U + F\delta b = V\delta Q$. The final result for F is the same, since at constant V we have $\partial U/\partial b = (V^2/2)\partial C/\partial b$ and $V\partial Q/\partial b = V^2 \partial C/\partial b$.

10.9   Setting $\varepsilon_r = 1$ in the gas bubble, an analysis similar to that in §10.6 shows that the field in the bubble is $3E_0 \varepsilon_r/(2\varepsilon_r + 1) = 11$ MV m^{-1}.

Chapter 11

11.1   A tangential electric field E acts on the electron, where $2\pi r E = \mathscr{E} = -d(\pi r^2 B)/dt = -\pi r^2 \, dB/dt$. Hence $m_e r(d\omega/dt) = -eE = \tfrac{1}{2}er(dB/dt)$, which gives $\Delta\omega = eB/2m_e$.

Also $\Delta(m_e r\omega^2) \approx 2m_e r\omega\Delta\omega = er\omega B = evB$.

The orbital moment of the electron $m = -er^2\omega/2$ (from equation (9.24)), so that $\Delta m = -e^2 r^2 B/4m_e$.

11.2 The magnetic moment of the atoms in a field B will be $m = -Ze^2\overline{r^2}B/4m_e$, and $M = Nm$. From equations (11.11), (11.12) and (11.13), $M = \chi_m H = \chi_m B/\mu_0(1+\chi_m) \approx \chi_m B/\mu_0$, leading to $\chi_m = -NZe^2\overline{r^2}/4\mu_0 m_e = 1.1 \times 10^{-6}Z$.

11.3 1.78×10^{-6}.

11.4 $\nabla \cdot \mathbf{M} = 0$ at interior points. Integrating $-\nabla\cdot\mathbf{M}$ through the surface (cf. §10.4) gives a 'magnetic surface charge density' $\mathbf{M}\cdot\mathbf{n}$. The solution parallels that to Problem 10.5. Inside the sphere, $\Phi_m = (M/3)r\cos\theta$, $\mathbf{H} = -\frac{1}{3}\mathbf{M}$, and $\mathbf{B} = \mu_0(\mathbf{H}+\mathbf{M}) = \frac{2}{3}\mu_0\mathbf{M}$.

11.5 The magnetic surface charges due to adjacent domains cancel, so there are no sources of $\mathbf{H}$.

11.6 Applied to a circular path through the ring, equation (11.10) gives $2\pi RH = NI$. Hence $B = \mu_r\mu_0 H = \mu_r\mu_0 NI/2\pi R$.

With a gap, $(2\pi R-b)H+bH_{gap} = NI$. Since B is continuous, $H_{ring} = \mu_r\mu_0 B$, $H_{gap} = \mu_0 B$, and the result follows.

Chapter 12

12.1 Comparing equation (12.6) with equation (6.1), the results of Chapter 6 are modified by $c \to c/\sqrt{(\varepsilon_r\mu_r)}$. In particular, in a plane wave $c|\mathbf{B}| = \sqrt{(\varepsilon_r\mu_r)}|\mathbf{E}|$. Equation (12.9) gives $U = \frac{1}{2}(\varepsilon_0\varepsilon_r|\mathbf{E}|^2 + |\mathbf{B}|^2/\mu_0\mu_r)$, and the results follow.

12.2 $\omega_0 = 1.59 \times 10^{16}$ s^{-1} ($\lambda_0 = 1190$ Å, $\hbar\omega_0 = 10.5$ eV).

12.3 24 m; 32 m ($K \sim 3 \times 10^{-4}$).

12.4
$$\mathbf{E}_r = \frac{1-n-i\kappa}{1+n+i\kappa}\mathbf{E}_i = \frac{1-n^2-\kappa^2-2i\kappa}{(1+n)^2+\kappa^2}\mathbf{E}_i.$$

Hence $\mathbf{E}_r = \sqrt{R}e^{i\phi}\mathbf{E}_i$, where $\tan\phi = 2\kappa/(n^2+\kappa^2-1)$.

12.5 Equation (12.24) easily generalises to the case of two media with refractive indices ε_1 and ε_2. Then

$$R_{12} = \left|\frac{\sqrt{\varepsilon_1}-\sqrt{\varepsilon_2}}{\sqrt{\varepsilon_1}+\sqrt{\varepsilon_2}}\right|^2,$$

which gives $R_{12} = R_{21}$.

12.6 The result uses Problem 12.5. Under the given conditions, the effect of multiple reflections is negligible.

12.7 Using $\nabla \times \mathbf{E} + \partial\mathbf{B}/\partial t = 0$, and $\partial\mathbf{B}/\partial t = -i\omega\mathbf{B}$,

$$\mathbf{B} = \begin{cases} \omega^{-1}[(-k_1,0,k_2)E_i e^{ik_1 z} + (k_1,0,k_2)E_r e^{-ik_1 z}]e^{ik_2 x - i\omega t} & \text{for } z < 0. \\ \omega^{-1}(-k_3,0,k_2)E_t e^{ik_3 z}e^{ik_2 x - i\omega t}, & \text{for } z > 0. \end{cases}$$

The dispersion relations are:

$$z < 0, \quad \omega^2\varepsilon_1/c^2 = k_2^2+k_1^2,$$
$$z > 0, \quad \omega^2\varepsilon_2/c^2 = k_2^2+k_3^2.$$

Hence

$$\frac{\varepsilon_2}{\varepsilon_1} = \frac{k_2^2+k_3^2}{k_2^2+k_1^2} = \frac{1+\cot^2\theta_t}{1+\cot^2\theta_i} = \frac{\sin^2\theta_i}{\sin^2\theta_t},$$

giving $\sqrt{\varepsilon_1}\sin\theta_i = \sqrt{\varepsilon_2}\sin\theta_t$.

The boundary conditions at $z = 0$ are satisfied for all x, t if

$$E_i + E_r = E_t \quad (E_\parallel, B_\perp \text{ continuous}),$$
$$k_1(E_i - E_r) = k_3 E_t \quad (B_\parallel \text{ continuous}),$$

giving

$$E_t = \frac{2E_i}{1 + k_3/k_1} = \frac{2E_i}{1 + \tan\theta_i/\tan\theta_t},$$

$$E_r = \frac{k_1 - k_3}{k_1 + k_3} E_i = \frac{\tan\theta_t - \tan\theta_i}{\tan\theta_t + \tan\theta_i} E_i.$$

12.8   From the dispersion relations,

$$\varepsilon_2/\varepsilon_1 = (1 + k_3^2/k_2^2)\sin^2\theta_i.$$

If $\varepsilon_1 \sin^2\theta_i > \varepsilon_2$, k_3^2 must be negative, $k_3^2 = -p^2$ say, where $p > 0$. Then $k_3 = \pm ip$. We must take $k_3 = ip$, so that the fields are finite as $z \to \infty$, and $e^{ik_3 z} = e^{-pz}$. Also $E_r = (k_1 - ip) E_i/(k_1 + ip)$, so that $|E_r| = |E_i|$, and all the energy is reflected.

12.9   An analysis similar to Problem 12.7, but using $\nabla \times \mathbf{B} - (\varepsilon_r/c^2)\partial\mathbf{E}/\partial t = 0$, gives

$$B_r = \frac{k_1/\varepsilon_1 - k_3/\varepsilon_2}{k_1/\varepsilon_1 + k_3/\varepsilon_2} B_i.$$

$B_r = 0$ if $k_3/k_1 = \varepsilon_2/\varepsilon_1$. From the dispersion relation, $k_2^2 + k_1^2 = (\varepsilon_1/\varepsilon_2)(k_2^2 + k_3^2)$. Since $\tan\theta_i = k_2/k_1$, this gives

$$\frac{k_3^2}{k_1^2} = \tan^2\theta_i\left(\frac{\varepsilon_2}{\varepsilon_1} - 1\right) + \frac{\varepsilon_2}{\varepsilon_1},$$

and the result follows.

Chapter 13

13.1   For copper at room temperature, $\sigma = 6 \times 10^7 \ \Omega^{-1} m^{-1}$, $\sigma/\varepsilon_0 \approx 7 \times 10^{18} \ s^{-1}$.

13.2   (a) For an estimate, take the collision times of the Na^+ and Cl^- ions equal. $\sigma \approx Ne^2\tau/M_{Na} + Ne^2\tau/M_{Cl}$ gives $\tau \sim 1.5 \times 10^{-14}$ s. $\omega\tau \ll 1$ if $\omega \ll 10^{14} \ s^{-1}$.
      (b) $\delta = [225/(n/Hz)^{1/2}]$ m.

13.3 $J(z) = \sigma_0 E(z) = \sigma_0 E_0 e^{-z/\delta} e^{i(z/\delta - \omega t)}$.   Total   current $I \approx 2\pi a \int J(z)\,dz = \text{Re}\,[2\pi r\sigma_0 E_0 \delta e^{-i\omega t}/(1-i)]$.   Hence $I = (2\pi a\sigma_0 E_0 \delta/\sqrt{2})\cos(\omega t - \pi/4) = I_0 \cos(\omega t - \pi/4)$.

Joule heating per unit length $\approx 2\pi a \int (\overline{J^2}/\sigma_0)\,dz$

$$= 2\pi a\sigma_0(E_0^2/2)\int e^{-2z/\delta}\,dz$$

$$= 2\pi a\sigma_0 E_0^2 \delta/4.$$

The result follows, using $I_0 = 2\pi a\sigma_0 E_0 \delta/\sqrt{2}$.

13.4   Taking $\varepsilon_r(\omega) = 1 - \omega_p^2/\omega^2$, the dispersion relation $\omega^2 = (c^2/\varepsilon_r)k^2$ yields $\omega^2 = \omega_p^2 + c^2k^2$.

$$v_{ph} = \omega/k = c[1 - (\omega_p/\omega)^2]^{-\frac{1}{2}},$$
$$v_g = d\omega/dk = c[1 - (\omega_p/\omega)^2]^{\frac{1}{2}}.$$

13.5 $\omega_p^2 = Ne^2/\varepsilon_0 m_e$ (equation (13.17)).

$$\omega \gg \omega_p = 5.6 \times 10^8 \text{ s}^{-1}.$$

13.6   The group velocity is $v_g = c[1-(\omega_p/\omega)^2]^{1/2}$. Suppose the distance is D. Then the delay time is

$$T = \frac{D}{c}\left[\frac{1}{[1-(\omega_p/\omega_1)^2]^{1/2}} - \frac{1}{[1-(\omega_p/\omega_2)^2]^{1/2}}\right]$$

$$\approx \frac{D\omega_p^2}{2c}\left[\frac{1}{\omega_1^2} - \frac{1}{\omega_2^2}\right], \quad \text{for } \omega \gg \omega_p.$$

Using $\omega_p^2 = Ne^2/\varepsilon_0 m_e$ gives $DN = 7.5 \times 10^{22}$ m^{-2}.

13.7 $\omega_p \approx 9.1 \times 10^{15}$ s^{-1} gives $N \approx 2.6 \times 10^{28}$ m^{-3}.

13.8
$$\varepsilon_r(\omega) = 1 + i\sigma_0/\varepsilon_0\,\omega(1-i\omega\tau)$$
$$= 1 - \sigma_0/\varepsilon_0\,\omega^2\tau(1+i/\omega\tau)$$
$$\approx 1 - \omega_p^2/\omega^2 + i\omega_p^2/\omega^3\tau,$$

expanding by the binomial theorem for $\omega\tau \gg 1$. For $\omega > \omega_p$, $[\varepsilon_r(\omega)]^{1/2} \approx [1-(\omega_p/\omega)^2]^{1/2} + i\omega_p^2/2\omega^3\tau[1-(\omega_p/\omega)^2]^{1/2}$. The result follows since $\varepsilon_r^{1/2} = n + i\kappa$.

Attenuation length for sodium = 33 μm.

13.9   The continuity equation is $\partial\rho/\partial t + \nabla\cdot\mathbf{J} = 0$. Using $\mathbf{J} = \sigma(\omega)\,\mathbf{E}$, $\nabla\cdot\mathbf{E} = \rho/\varepsilon_0$ gives

$$\partial\rho/\partial t + \sigma(\omega)\,\rho/\varepsilon_0 = 0.$$

This is satisfied by $\rho = \rho_0\,e^{-i\omega t}$ if $i\omega = \sigma(\omega)/\varepsilon_0$, which gives $\omega = \omega_p$.

Chapter 14

14.1   From Fig. 14.2, $B_c = 0.057$ T at 4 K. By Ampères law, the maximum field when the current is I is at the surface of the wire, and is given by $2\pi rB = \mu_0 I$. Hence when $B = B_c$, $I = 285$ A.

14.2   Inside a uniformly magnetised sphere, $\mathbf{B} = \frac{2}{3}\mu_0\mathbf{M}$ (Problem 11.4). Thus an external field $\mathbf{B}_{ext}$ is cancelled if $\mu_0\mathbf{M} = -\frac{3}{2}\mathbf{B}_{ext}$.

The magnetic potential outside the sphere (see Problem 11.4) is

$$\Phi_m = -(B_{ext}/\mu_0)\,r\cos\theta + Ma^3\cos\theta/3r^2.$$

The field at the sphere is parallel to the surface, and is

$$B = \mu_0 H = \mu_0(-\partial\Phi_m/r\,\partial\theta)_{r=a}$$
$$= (B_{ext} - \tfrac{1}{3}\mu_0 M)\sin\theta.$$

This is a maximum at $\theta = \pi/2$, $B = \frac{3}{2}B_{ext}$; and exceeds B_c if $B_{ext} > \frac{2}{3}B_c$.

14.3   If the distance is a, each triangle has area $\sqrt{3}a^2/4$. Each triangle has $\frac{1}{2}$ (flux tube) associated with it. Consider N tubes. The total flux is $N\Phi_0$, and they fill an area $N\sqrt{3}a^2/2$. Hence the mean magnetic field $B = 2\Phi_0/\sqrt{3}a^2$.

Now $B = B_{ext} + \mu_0 M \approx B_{ext}$ (since $\mu_0 M$ is small for $B_{ext} = 0.12$T), so $B_{ext} \approx 2\Phi_0/\sqrt{3}a^2$, giving $a = 0.14$ μm.

Suppose there is a compressional energy $e(a)$ per unit length between adjacent tubes. Each tube has an associated energy $\varepsilon + 3e(a)$, and no more tubes enter the sample when $B_{ext}\Phi_0/\mu_0 = \varepsilon + 3e(a)$, from §14.6. Hence the force between the tubes

$$f = -\frac{de}{da} = -\frac{\Phi_0}{3\mu_0}\frac{dB_{ext}}{da} \approx \frac{aB_{ext}^2}{\sqrt{3\mu_0}} = 9.2 \times 10^{-4} \text{ N m}^{-1}.$$

14.4   The flux through the solenoid is $\mathscr{F} = \pi R^2 B$. The magnetic energy per unit length is $\varepsilon = (B^2/2\mu_0)\pi R^2 = \mathscr{F}^2/2\mu_0 \pi R^2$. Hence if $R \to R + dR$ at constant $\mathscr{F}$, $d\varepsilon = -(\mathscr{F}^2/\mu_0 \pi R^3)dR = -(\pi R B^2/\mu_0)dR$. If p is the equivalent pressure, $d\varepsilon + 2\pi R p\, dR = 0$, giving $p = B^2/2\mu_0$. For B = 10 T, $p = 393$ atm.

14.5   The self inductance of the ring is $L \sim \mu_0 b \ln(b/a) = 5.8 \times 10^{-8}$ H (Problem 9.15).

The resistance of the circuit is $< 2b\rho/a^2 = 7.2 \times 10^{-19}\ \Omega$, giving a time constant $\tau = L/R > 8 \times 10^{10}$ s ~ 2500 yr.

Chapter 15

15.1 $\dfrac{dV}{dz} = -\dfrac{1}{\varepsilon_0}\displaystyle\int_{-\infty}^{z}\rho(z'')\,dz''; \quad V(z) = -\dfrac{1}{\varepsilon_0}\displaystyle\int_{-\infty}^{z}dz'\int_{-\infty}^{z'}\rho(z'')\,dz''.$

Integrating the z' integral by parts,

$$V(z) = -\frac{1}{\varepsilon_0}\left[z\int_{-\infty}^{z}\rho(z'')\,dz'' - \int_{-\infty}^{z}z'\rho(z')\,dz'\right].$$

Since $\int_{-\infty}^{\infty}\rho(z'')\,dz'' = 0$, $\varepsilon_0 V_{21} = \int_{-\infty}^{\infty}z'\rho(z')\,dz'$.

15.2   When the capacitor plate (2) carries charge Q, $Q/(-e)$ electrons have been transferred from plate (1) to (2). The net work done at the surfaces is $[Q/(-e)](W_1 - W_2)$, and $Q = CV$. The term $CV^2/2$ is the energy stored in the field.

15.3   The potential between the plates is $V = (5.65 - 3.4)$ V. Since $C = \varepsilon_0 A/d$, $Q = CV = \varepsilon_0 AV/d$. Hence a charge $\varepsilon_0 AV(1/d_1 - 1/d_2) = 4.98 \times 10^{-11}$ C is transferred to the platinum plate.

15.4 $\nabla^2\Phi = 0$, by substitution.

For $y = 0$, $x > 0$, $\Phi = \Phi_{Pt} = $ constant.

For $y = 0$, $x < 0$, $\Phi = \Phi_{Th} = $ constant $+ 0.72\pi$ V.

At the platinum surface the electric field is normal and $E_y = -(\partial\Phi/\partial y)_{y-0}$ $= -0.72/x$ V. The charge density σ is given by $\sigma = \varepsilon_0 E_y$, and $N = \sigma/(-e)$.

15.5   From equation (15.4), $d = [2\varepsilon_0 V_0/N_0 e]^{1/2} = 1050$ Å. The additional voltage draws electrons from the depletion layer, so that d becomes $d_1 = [2\varepsilon_0(V_0 + V_b)/N_0 e]^{1/2}$. The surface charge on the metal is the negative of the charge in the depletion layer. Hence

$$\sigma = -eN_0 d_1 = -[2\varepsilon_0 N_0 e(V_0 + V_b)]^{\frac{1}{2}}$$
$$-d\sigma/dV_b = [\varepsilon_0 N_0 e/2(V_0 + V_b)]^{\frac{1}{2}} = 59\ \mu\text{F m}^{-2}$$

for $V_b = 1$ V.

15.6   Note that each SO_4^{2-} consumed produces *one* net electron.

9.2×10^{23} electrons $\to 1.47 \times 10^5$ C = 41 A hr.

Energy = 2 eV per electron.

Chapter 16

16.1 $\chi(\mathbf{r}, t)$ must satisfy $\nabla^2\chi - \partial^2\chi/c^2\,\partial t^2 = -F(\mathbf{r}, t)$, which has the given solution (cf. equation (16.15)).

16.2 (a) $E = \frac{1}{2}m_e\dot{\mathbf{r}}^2 + \frac{1}{2}m_e\omega_0^2\mathbf{r}^2 = \frac{1}{2}m_e\omega_0^2 r_0^2$.
(b) This follows from equation (16.22), since $\mathbf{p} = e\mathbf{r}_0\cos\omega_0 t$.
(c) We need $r_0/c \ll 1/\omega_0$, i.e., $r_0 \ll c/\omega_0 = \lambda/2\pi$.
(d) Use equation (16.26), noting that the time-averaged value of $|\ddot{\mathbf{p}}|^2 = \omega_0^4 e^2 r_0^2/2$, together with (a) above.

16.3 Take $\mathbf{p} = (p_0\cos\omega t, p_0\sin\omega t, 0)$. Then $\ddot{\mathbf{p}} = -\omega^2(p_0\cos\omega t, p_0\sin\omega t, 0)$, $|\ddot{\mathbf{p}}|^2 = \omega^4 p_0^2$. Hence
$$\frac{d\mathscr{E}}{dt} = \frac{2}{3}\frac{\omega^4 p_0^2}{4\pi\varepsilon_0 c^3}.$$

16.4 $m = cp$ follows on transforming the electric dipole formula (3.5) into that of the magnetic dipole (9.13). Making the transformation on the time-dependent dipole leaves the Poynting vector unchanged, so that one might expect
$$\frac{d\mathscr{E}}{dt} = \frac{2}{3}\frac{1}{4\pi\varepsilon_0 c^5}|\dddot{\mathbf{m}}|^2 = \frac{2}{3}\frac{\mu_0}{4\pi c^3}|\dddot{\mathbf{m}}|^2.$$
This is correct.

16.5 $\mathbf{m} = m_0(\sin\alpha\cos\omega_0 t, \sin\alpha\sin\omega_0 t, \cos\alpha)$, giving $\ddot{\mathbf{m}} = -m_0\omega_0^2\sin\alpha(\cos\omega_0 t, \sin\omega_0 t, 0)$. The result then follows from Problem 16.4.
Rotational energy $\mathscr{E} = \frac{1}{2}I\omega_0^2 \approx \frac{1}{2}I(2\pi/T)^2$, where $T =$ length of day. Hence $d\mathscr{E}/dt = -[I(2\pi)^2/T^3]\,dT/dt$. Equating this to the rate of radiation of energy, $dT/dt = 3.25 \times 10^{-30} = 2.8 \times 10^{-25}$ seconds per day.

Chapter 17

17.1 The power $P(\theta)\,d\theta$ radiated in a range $\theta, \theta + d\theta$ is proportional to $\sin^3\theta\,d\theta$ (cf. equation (16.26)). Then $\int_0^\alpha \sin^3\theta\,d\theta = \frac{1}{2}\int_0^{\pi/2}\sin^3\theta\,d\theta$ implies $\cos^3\alpha - 3\cos\alpha + 1 = 0$. Hence $\alpha \approx 70°$, $90° - \alpha \approx 20°$.

17.2 In computing $\nabla \times \mathbf{A}$ at large distances, only the term $\cos\omega(t - r/c)$ need be differentiated, since all other terms are smaller by powers of (λ/r) and are negligible.

17.3 Power $= \frac{1}{2}R_{\lambda/2}I_0^2$, where $R_{\lambda/2} = 73.1\ \Omega$. Hence $I_0 = 5.23$ A.
At $\theta = \pi/2$, power $(\mu_0/\varepsilon_0)^{1/2}I_0^2\,d\theta/4\pi$ is radiated in a range $d\theta$ (Problem 17.2). But for large r, $P(\theta)\,d\theta = 2\pi r^2\,d\theta|\mathbf{N}| = 2\pi r^2 c(B_0^2/2\mu_0)\,d\theta$. Hence $B_0 = \mu_0 I_0/2\pi r$.
At $r = 1$ km, $B_0 = 1.05 \times 10^{-10}$ T, and $E_0 = cB_0 = 0.0314$ V m^{-1}.

17.4 Take the dipoles at $\mathbf{R}_1 = (0, 0, d/2)$, $\mathbf{R}_2 = (0, 0, -d/2)$
$$\hat{\mathbf{r}} = (\sin\theta\cos\phi, \sin\theta\sin\phi, \cos\theta).$$
$$P(\theta) = (\text{constant})\sin^3\theta|\exp(i\pi d\cos\theta/\lambda) + \exp(-i\pi d\cos\theta/\lambda)|^2,$$
using equation (17.7). Hence the result.
Choosing $d = \lambda/2$, $\cos^2(\frac{1}{2}\pi\cos\theta) \approx 1$ for $\theta \approx \pi/2$, and $\approx \sin^2(\pi\theta^2/4) \approx \pi^2\theta^4/16$ for $\theta \approx 0$.

17.5 Take $\mathbf{R}_1 = (-\lambda/8, 0, 0)$, $\phi_1 = \pi/4$; $\mathbf{R}_2 = (\lambda/8, 0, 0)$, $\phi_2 = -\pi/4$. In the horizontal plane $\theta = \pi/2$, $\hat{\mathbf{r}} = (\cos\phi, \sin\phi, 0)$. Hence
$$P(\phi) = (\text{constant})\cos^2\left[\tfrac{1}{4}\pi(1 - \cos\phi)\right].$$

17.6   Consider the second dipole to be at $(D, 0, 0)$.

$$\mathbf{N}(\theta, \phi) = \mathbf{N}_0(\theta, \phi) |1 - \exp[i(2\pi D/\lambda) \sin\theta \cos\phi]|^2$$
$$\approx \mathbf{N}_0(\theta, \phi)(2\pi D/\lambda)^2 \sin^2\theta \cos^2\phi.$$

In equation (16.26), the angular integral $\sin^3\theta \, d\theta \, d\phi$ becomes $\sin^5\theta \cos^2\phi$ $d\theta \, d\phi$, and the result follows.

At 50 Hz, $\lambda = 6000$ km. Hence the cable can be considered to radiate as two simple dipoles, and $R_2/R_1 \approx 4.4 \times 10^{-13}$.

17.7   Classically, $m_e r\omega^2 = e^2/4\pi\varepsilon_0 r^2$. Kinetic energy $= \frac{1}{2}m_e r^2\omega^2 = \frac{1}{2}e^2/4\pi\varepsilon_0 r$, potential energy $= -e^2/4\pi\varepsilon_0 r$. Adding, total energy $E = -\frac{1}{2}e^2/4\pi\varepsilon_0 r$.

Larmor's formula gives

$$\frac{dE}{dt} = -\frac{2}{3}\left(\frac{e^2}{4\pi\varepsilon_0}\right)\frac{r^2\omega^4}{c^3}.$$

Then

$$\frac{dr}{dt} = \frac{dr}{dE}\frac{dE}{dt} = -\frac{4}{3}\frac{r^4\omega^4}{c^3} = -\frac{4}{3}\left(\frac{e^2}{4\pi\varepsilon_0}\right)^2\frac{1}{m_e^2 c^3 r^2}.$$

Integrating,

$$t_2 - t_1 = \frac{m_e^2 c^3}{4(e^2/4\pi\varepsilon_0)^2}(r_1^3 - r_2^3).$$

(a) 8.7×10^{-5} s.

(b) The electron is emitting radiation at its rotation frequency. Eliminating r, $E = -\frac{1}{2}(e^2/4\pi\varepsilon_0)^{2/3} m_e^{1/3}\omega^{2/3}$. Hence the result.

17.8   The positron has tunnelled through the Coulomb barrier at a distance r_0 given by $Ze^2/4\pi\varepsilon_0 r_0 = E_e$. Its acceleration thereafter is given by $m_e a = Ze^2/4\pi\varepsilon_0 r^2$, and Larmor's formula gives

$$\frac{dE_\gamma}{dt} = \frac{2}{3}\left(\frac{e^2}{4\pi\varepsilon_0}\right)\left(\frac{Ze^2}{4\pi\varepsilon_0}\right)^2\frac{1}{m_e^2 c^3 r^4}.$$

Now $dE_\gamma/dr = (dE_\gamma/dt)/(dr/dt)$, and (dr/dt) is given approximately by the energy equation

$$\tfrac{1}{2}m_e(dr/dt)^2 \approx E_e - Ze^2/4\pi\varepsilon_0 r.$$

The given integral is then obtained after the change of variable $r = (Ze^2/4\pi\varepsilon_0 E_e)x$. Note $Z = 8$. The integral ≈ 1.07, and $E_\gamma = 5$ keV $\ll E_e$. (This is only an estimate: our non-relativistic treatment is here only a rough approximation.)

17.10   If the damping is small, $\mathbf{r} \approx \mathbf{r}_0 \cos\omega t$, and $m\tau\dddot{\mathbf{r}} \approx -\omega_0^2 m\tau\dot{\mathbf{r}}$ in equation (17.9). (See also Problem 16.2.)

17.11   For $\lambda \gg a$, we can take the sphere to be uniformly polarised at any instant. From §10.6, the dipole moment in an electric field $E_0 \sin\omega t$ is

$$\mathbf{p}(t) = 4\pi a^3\varepsilon_0\left(\frac{\varepsilon_r - 1}{\varepsilon_r + 2}\right)E_0 \sin\omega t.$$

The mean rate of radiation is, from equation (16.26),

$$\frac{2}{3}\frac{\varepsilon_0}{4\pi c^3}(4\pi a^3)^2\left(\frac{\varepsilon_r - 1}{\varepsilon_r + 2}\right)^2\frac{E_0^2\omega^4}{2}.$$

The mean incident flux is $\varepsilon_0 c E_0^2/2$, and the result follows from the definition of cross-section.

Chapter 18

18.1   If in the cable $I(z,t) = I_i(z-ct) + I_r(z+ct)$, then $V(z,t) = Z_0[I_i(z-ct) - I_r(z+ct)]$. Suppose the cable is terminated with the resistive load at $z = 0$. Then $V = IR$ at $z = 0$, i.e., $Z_0[I_i(-ct) - I_r(ct)] = R[I_i(-ct) + I_r(ct)]$, so that $I_r(ct) = I_i(-ct)(Z_0 - R)/(Z_0 + R)$, and there is a reflected wave

$$I_r(z+ct) = I_i(-z-ct)(Z_0 - R)/(Z_0 + R).$$

18.2   Let $I(z,t) = [e^{2\pi iz/\lambda} + B e^{-2\pi iz/\lambda}]e^{-i\omega t}$, then $V(z,t) = Z_0[e^{2\pi iz/\lambda} - B e^{-2\pi iz/\lambda}]e^{-i\omega t}$.
(a)   If $I(\lambda/4, t) = 0$, $B = 1$ and $V(0,t) = 0$, $Z_i = 0$.
(b)   If $V(\lambda/4, t) = 0$, $B = -1$ and $I(0,t) = 0$, $Z_i = \infty$.
In general $Z_i = Z_0(1 - B)/(1 + B)$ and $Z = Z_0(1 + B)/(1 - B)$, giving $ZZ_i = Z_0^2$. This result is useful in impedance matching.

18.3   The attenuation length is

$$d = \left(\frac{2\sigma}{\omega \varepsilon_0 \varepsilon_r}\right)^{1/2} \frac{b \ln x}{1+x} \text{ where } x = b/a.$$

For fixed b, this is a minimum where

$$\frac{d}{dx}\left(\frac{\ln x}{1+x}\right) = 0,$$

i.e., $1 + x = x \ln x$.

18.4   In the geometry of the figure, between the strips take

$$\mathbf{E} = [V(z,t)/h](1,0,0)$$
$$\mathbf{B} = [\mu_0 I(z,t)/d](0,1,0).$$

Then $\nabla \cdot \mathbf{E} = 0$ and $\nabla \cdot \mathbf{B} = 0$.
$\nabla \times \mathbf{E} + \partial \mathbf{B}/\partial t = 0$ yields

$$\partial V/\partial z = -(\mu_0 h/d)\partial I/\partial t.$$

$\nabla \times \mathbf{B} - (\varepsilon_r/c^2)\partial \mathbf{E}/\partial t = 0$ yields $-\partial I/\partial z = (\varepsilon_r d/c^2\mu_0 h)\partial V/\partial t$.
The solution follows §18.1, but with velocity of propagation $c/\sqrt{\varepsilon_r}$, and $Z_0 = (c\mu_0/\sqrt{\varepsilon_r})h/d = (\mu_0/\varepsilon_0 \varepsilon_r)^{1/2}h/d$.

18.5   By the formula of trigonometry,

$$\sin a \cos b = \tfrac{1}{2}[\sin(a+b) + \sin(a-b)],$$

the standing waves are a special case of the general solution (18.6) and (18.7), with

$$I_+ = \tfrac{1}{2}A\sin\frac{2\pi}{\lambda}(z-ct), \quad I_- = \tfrac{1}{2}A\sin\frac{2\pi}{\lambda}(z+ct).$$

The boundary condition, $I = 0$ at $z = 0$, is satisfied.
For $-l/2 < z < 0$, $2\pi z/\lambda \ll 1$. Writing

$$z' = z+l/2, \quad I \approx A(2\pi/\lambda)(z'-l/2)\cos\omega t, \quad V \approx -Z_0 A\sin\omega t,$$

where $\omega = 2\pi c/\lambda$. At $z' = 0$, the impedance is imaginary since V and I are $\pi/2$ out of phase. $|V_0|/|I_0| = Z_0(\lambda/\pi l)$.

18.6 $\omega^2 = \omega_{mn}^2 + c^2 k^2$,

$$v_{phase} = \omega/k = c[1 - (\omega_{mn}/\omega)^2]^{-1/2},$$

$$v_{group} = d\omega/dk = c[1 - (\omega_{mn}/\omega)^2]^{1/2} = c^2/v_{phase}.$$

18.7   From equations (18.16) and (18.17), the time average of the Poynting vector $\mathbf{E} \times \mathbf{B}/\mu_0$ in the guide is $[0, 0, (kE_0^2/2\mu_0\omega)\sin^2(\pi x/a)]$. Integrating over the cross-section of the guide gives the result.

For $\omega = 6\pi \times 10^9$ s^{-1}, $k = 34.7$ m^{-1} (TE$_{1,0}$ mode), giving maximum power = 3.5 MW.

$\omega_{min} = \pi c/a$, $v_{min} = \omega_{min}/2\pi = c/2a = 2.5$ GHz.

For $\omega < \omega_{min}$, $k^2 = -K^2$ where $K = (\omega_{min}^2 - \omega^2)^{1/2}/c$. Attenuation length $= 1/2K = 6.8$ cm. The next mode has frequency $\frac{3}{2}v_{min}$, so $v_{min} < v < \frac{3}{2}v_{min}$.

18.8   160 m.

Chapter 19

19.1   Under the Galilean transformation, the Lorentz equation $d\mathbf{p}/dt = Q(\mathbf{E} + \mathbf{v} \times \mathbf{B})$ becomes $d\mathbf{p}'/dt' = Q[\mathbf{E}' + (\mathbf{v} - \mathbf{u}) \times \mathbf{B}'] = Q(\mathbf{E} + \mathbf{v} \times \mathbf{B})$. This holds for all $\mathbf{v}$ if $\mathbf{B}' = \mathbf{B}$; then $\mathbf{E}' = \mathbf{E} + \mathbf{u} \times \mathbf{B}$.

19.2   From equation (19.5), $d\mathscr{E}/dt = 0$ if $\mathbf{E} = 0$. Hence $\mathscr{E}$ and γ are constant. Equation (19.4) then becomes $\gamma m\,d\mathbf{v}/dt = (-e)\mathbf{v} \times \mathbf{B}$, which is of the same form as Problem 4.1 but with $\omega = eB/\gamma m = eBc^2/\mathscr{E}$.

19.5   For a charge ΔQ at x_n at time t, the formula (19.7) generalises to

$$B_x = 0, \quad B_y = 0, \quad B_z(t) = \frac{\mu_0}{4\pi}\frac{\gamma u a \Delta Q}{[a^2 + \gamma^2(ut + x_n)^2]^{3/2}}.$$

The total field is

$$B_z(t) = \frac{\mu_0}{4\pi}\sum_{n=-\infty}^{\infty}\frac{\gamma u a \Delta Q}{[a^2 + \gamma^2(ut + nu\Delta t)^2]^{3/2}}$$

$$\rightarrow \frac{\mu_0}{4\pi}\int_{-\infty}^{\infty}\frac{\gamma u a \Delta Q\,dn}{[a^2 + \gamma^2(ut + nu\Delta t)^2]^{3/2}}.$$

The change of variable $\alpha = \gamma(ut + nu\Delta t)/a$ gives

$$B_z(t) = \frac{\mu_0 I}{4\pi a}\int_{-\infty}^{\infty}\frac{d\alpha}{[1 + \alpha^2]^{3/2}} = \frac{\mu_0 I}{2\pi a}.$$

Index